12/6/93

PHYSIOLOGICAL ANIMAL EC

PHYSIOLOGICAL ANIMAL ECOLOGY

GIDEON N. LOUW, Ph.D. (Cornell)

Zoological Institute
University of Zürich

Longman
Scientific &
Technical

Longman Scientific and Technical
Longman Group UK Limited
Longman House, Burnt Mill, Harlow
Essex CM20 2JE, England
and Associated Companies throughout the world.

Copublished in the United States with
John Wiley & Sons, Inc., 605 Third Avenue, New York, NY 10158

First published in 1993

British Library Cataloguing in Publication Data
A catalogue record for this book is available from the British Library.

ISBN 0-582-05922-4

Library of Congress Cataloging-in-Publication Data

Louw, Gideon, 1930–
  Physiological animal ecology/Gideon N. Louw.
    p.   cm.
  Includes bibliographical references and index.
  1. Animal ecophysiology. I. Title.
QP82.L668 1992
591.5—dc20

Set in Baskerville 10/12 pt
Printed in Hong Kong
WLEE/01

Contents

Preface

This volume is intended for undergraduate students. It is especially targeted at students of animal ecology to provide a bridge between their knowledge of ecology and their understanding of physiology. The author has, however, presupposed an elementary knowledge of the physical sciences, physiology, biochemistry and zoology, such as is contained in a good freshman biology course. Advanced sixth form scholars at British schools may also benefit from reading it, and graduate students wishing to enter a new field of interest should find it of value.

Several people have contributed importantly to the production of this volume and I take great pleasure in thanking them. First, Professor Rüdiger Wehner of Zürich University who provided the author with a visiting professorship, when the bulk of the volume was written. He also provided many ideas and much encouragement. Special thanks to Professor Duncan Mitchell who edited Chapters 1 and 2 with great care and generously provided many helpful suggestions and examples. Professor Neil Hadley is thanked for the free use of his library and lecture notes, as well as his advice and encouragement during our long association as colleagues and friends. Cornelia Jud prepared the first draft of the manuscript with customary Swiss efficiency and enthusiasm. Jenny Hale undertook the production of the final draft and I am most grateful for her devoted and cheerful assistance. Sincere thanks are also due to my publishers. I wish to thank Dr R. R. Arndt, President of the Foundation for Research Development who allowed me to use the facilities of the FRD to complete this volume. Finally my gratitude to Claire for her constant help and encouragement, provided in innumerable ways.

G.N. Louw

Acknowledgements

We are grateful to the following for permission to reproduce copyright material:

Academic Press Inc. and the author for fig. 3.2 (Harbourne, 1982); *American Journal of Physiology* for fig. 3.16 (Taylor *et al.*, 1970); American Society of Zoologists and the author for fig. 3.9 (Graham, 1990); Blackie Academic and Professional, an imprint of Chapman and Hall (publishers) Ltd for figs 2.20 (Phillips *et al.*, 1985); 4.5 (Goldsworthy *et al.*, 1981) and tables 2.5 (Rankin & Davenport, 1981), 3.1 and 3.2 (Brafield & Llewellyn, 1982); Butterworth-Heinemann (publishers) Ltd for fig. 3.10 and tables 3.7, 3.8 and 3.9 (Durnin & Passmore, 1967); Cambridge University Press and the authors for figs 1.11, 1.37 and 1.40, tables 2.4 and 3.3 (Schmidt-Nielsen, 1983), fig. 3.12 (Schmidt-Nielsen, 1984), fig. 4.4 (Austin & Short, 1972) and fig. 4.13 (Setchell, 1982); Cooper Ornithological Society and the authors for fig. 3.28 (Mugaas & King, 1981); Copeia and the authors for fig. 2.17 (Blaylock *et al.*, 1976); CRC Press Inc. for tables 2.2 and 2.3 (The CRC Handbook of Chemistry and Physics, 60th edition); Ecological Society of America and the authors for fig. 1.8 (Porter & Gates, 1969) and the author for table 3.4 (Nagy, 1987); W H Freeman and Company (publishers) Inc. for fig. 1.32 (Eckert & Randall, 1983); Liviana Editrice SpA and the author, Dr V H Shoemaker, for fig. 2.19 (Dejours *et al*, 1987); Oecologica and the authors for table 3.10 (Taigen *et al.*, 1982); Oxford University Press for fig. 1.5 (Schmidt-Nielsen, 1964); Scientific American for figs 4.11 and 4.12 (Crews & Garstaka, 1982); Tom Ulrich for plate 1; John Wiley and Sons Inc. for figs 3.21 and 3.22 (Hadley, 1985).

Whilst every effort has been made to trace the owners of copyright material, in a few cases this has proved impossible and we take this opportunity to offer our apologies to any copyright holders whose rights we may have unwittingly infringed.

INTRODUCTION—WHAT IS
PHYSIOLOGICAL ECOLOGY?

To best answer the above question, we should first briefly look at the recent history of both ecology and physiology. Had you been a student of zoology in the 1920s and 1930s, your curriculum would have consisted mostly of comparative anatomy and taxonomy. In many German and British universities this was often a test of endurance and memory rather than a process of intellectual enrichment. After World War II, zoology courses started to change rapidly. At first increasing attention was given to physiology, a subject hitherto taught mainly by medical schools. The physiology taught by zoology departments was, however, strongly comparative and sought, whenever possible, to provide additional support for Darwin's theory of evolution. These physiological studies were almost all carried out in the laboratory under artificial conditions. It was only in the early 1950s that a significant number of physiologists and zoologists began to study the physiological responses of animals in the field as well as in the laboratory. The major pioneers of this era were George (Bart) Bartholomew of UCLA, Knut Schmidt-Nielsen of Duke University and Pete Scholander of the Scripps Institute. At first these studies tended to concentrate on the physiological responses of animals to extreme environments such as deserts, the ocean depths and polar regions. They also asked 'how' questions such as 'How do camels survive without daily access to water?' or 'How are polar bears and seals able to survive immersion in polar seas?' Today these studies on the physiological adaptations of animals are usually grouped under the heading of *environmental physiology*.

Trailing slightly behind the exciting developments in environmental physiology came the almost explosive rise in popularity of ecology. Ecology developed from earlier simple field studies performed by amateurs, and known as natural history. It was at first very descriptive and also asked mostly 'how' questions such as 'How many animals occur in this area?' or 'How widely are these animals distributed?' Before long, however, these studies became far more rigorous with the introduction of advanced statistical methods, mathematical modelling and the critical testing of new concepts. Early in its history, ecologists started asking 'why' questions in addition to 'how' questions, such as 'Why are there far fewer species of animals

Plate 0.1 Physiological ecologists capturing seals in the Antarctic to study their blood chemistry before releasing them again. (Photo: courtesy J. D. Skinner.)

at higher latitudes than close to the Equator?' This pattern of questioning soon led to a strong involvement of evolutionary theory in ecology.

In spite of the important advances within ecology, many ecologists still today tend to treat their animals as black boxes and do not give sufficient attention to their unique physiological roles within an ecosystem. Conversely, physiologists tend to concentrate on physiological adaptations and 'how' questions, without sufficient regard to the broader ecological and evolutionary implications of their work. The present volume is therefore intended to provide students of animal ecology with a basic knowledge of the major physiological adaptations exhibited by animals to the environment, with the hope that this will enrich their understanding and insight into ecology. In this regard, the author is well aware of the debate surrounding even the use of the term *adaptation* and that some biologists question whether adaptations exist at all (see Gould and Lewontin, 1979). Highly theoretical views of this nature are best treated elsewhere, and my interpretation of the term adaptation is simply an observable physiological or behavioural characteristic which allows an organism to survive and reproduce in a particular environment, thereby contributing to its fitness.

The ultimate aim of physiological ecology then is to ask 'how' as well as 'why' questions, thereby making use of physiological data to gain a better understanding of the distribution, abundance and evolution of organisms. In this way we can deepen our knowledge of both theoretical and applied ecology. For example, when a physiological ecologist is confronted with the distribution of the largest bird on this planet, the ostrich, he/she would immediately be impressed by the ability of this species to survive under extreme desert conditions. This could prompt an investigation of the behaviour, nutrition, water turnover rates, and thermoregulation of these unusual birds both in the field under natural conditions as well as in the laboratory. From these studies it would be apparent that ostriches possess remarkable physiological attributes, such as the ability to expire air that is not saturated with water vapour as well as having efficient kidney function and behavioural responses which minimise heat gains in hot environments. Finally, however, the investigator will conclude that it is not a single dramatic physiological characteristic that ensures the survival of ostriches under desert conditions but rather a series of adaptations acting in concert, including the large size of the bird. Having now answered 'how' the animal survives, the investigator will move into the more speculative field of perhaps asking 'why'—why has a large flightless bird been so successful in the semi-arid ecosystems of Africa and the Middle East; why are these birds apparently not seriously affected by competition and predation; why have large flightless birds (ratites) evolved separately and successfully on three separate continents? In this way the investigator moves from basic physiology through traditional ecology and, if so inclined, ultimately to evolutionary theory.

In this book you will therefore learn how starfish reproduce, why desert beetles collect water by fog basking, how bears synthesise protein during their winter torpor, why bedouin goats are black, why humming-birds spend so much time resting between foraging bouts, why ostriches erect their feathers to keep cool, how marlin keep their brains warm and how cheetahs keep their brains cool while sprinting, to name but a few examples. Considerable attention has also been given to the more important techniques involved in the many studies that have been cited. This has been done in the belief that students gain a more critical and deeper appreciation of physiological phenomena when they understand the methods that have been used to obtain that knowledge.

The material is arranged in four major sections, namely temperature regulation, water relations, nutrition and energy, and reproduction and the environment, these being the major physiological systems, apart from the sensory system, which interface with the environment. Within each section, the discussion begins first with the physical principles involved in that particular topic. This is followed by the physiological principles and then by a description of actual case studies. Each section is briefly summarised under the title 'concluding remarks', when the author indulges in some mild speculation of an ecological and evolutionary nature. The natural flow from physics to evolution is therefore from hard to softer science. However, no attempt has been made to analyse the literature through the eyes of a modern theoretical evolutionist.

This volume consequently represents the current state of knowledge obtained from traditional studies in physiological ecology. There is, however, no doubt that physiological animal ecology is entering a new phase in which molecular biology, population genetics, chaos theory, systems theory and evolutionary theory will play an increasingly important role. The serious student of these new trends should consult Feder *et al.* (1987) in this regard. Nevertheless, if we trace the history of physiological animal ecology it is still firmly rooted in natural history and this text can perhaps best be described as modern natural history. It has no pretensions to great originality or profound theory, but wishes merely to impart the excitement of finding out more about how animals 'work' in their particular environments.

CHAPTER 1

TEMPERATURE AND THERMOREGULATION

1.1 Introduction

It would be naïve to try to select the 'most important' environmental factor affecting life when, for example, the absence of a few milligrams of any one of several trace elements from the diet of a large mammal will lead to severe illness and eventual death. Nevertheless, no one would deny that temperature has a profound effect on all life processes and so warrants careful study and analysis. Such study requires familiarity with the basic effects of temperature on chemical reaction rate and with the well-known Q_{10} effect.

Life is sustained by a complex of biochemical reactions and the molecules involved in these reactions possess kinetic energy, that is, they are in constant motion. If we raise the temperature of reacting molecules from 30 to 40 °C we increase their kinetic energy by about 3%, yet the reaction velocity may increase 200–300%. The reason for the disproportionate increase can best be explained by referring to Fig. 1.1. The feasibility of a reaction taking place is determined by the energy barrier to the reaction, known as the activation energy. According to the Arrhenius principle, a relatively small percentage change in the average kinetic energy of a population of molecules may result in a relatively large change in the fraction of molecules having energy greater than the activation energy. This phenomenon is the basis of the profound effect of temperature upon life processes.

It is customary to describe the effect of temperature on a life process in terms of the Q_{10}:

$$Q_{10} = \text{Velocity } (T + 10\,°C)/\text{Velocity } (T°C) \qquad [1.1]$$

Because the biochemical reaction velocity does not increase linearly with the temperature, Q_{10} is not a constant for a particular reaction, but itself varies with temperature. It is important, therefore, to specify the range of temperature over which a Q_{10} was measured.

The more chemical reactions upon which a living organism depends, the greater the effect of temperature on its functioning, and the human brain, with its multiple synaptic connections, is potentially the most temperature sensitive of all living tissues. Yet, if we look around any room in which 20–30

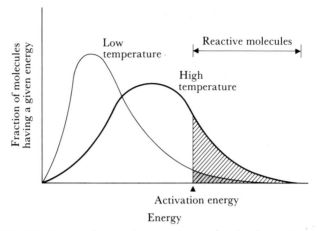

Figure 1.1   Distribution of energy in a population of molecules at a low and high temperature. Only those molecules having energy equal to or greater than the activation energy are reactive.

persons are gathered, we will very likely find one who can tune and play a string instrument, one who can solve a differential equation, perhaps one who can write some poetry; probably all will be able to use complex speech patterns for communication and exercise fine motor control over their voluntary muscles. It is not therefore surprising to find that human body temperatures are remarkably constant. It seems likely that the development of the impressive intellectual ability of humans was and still is dependent on very fine thermal control over the central nervous system. One only has to observe the deranged behaviour of humans in whom brain temperatures have even been slightly disturbed to realise the importance of a regulated brain temperature.

The development of the ability to regulate body temperature may have been a prerequisite for the development of complex nervous systems. Many other authors consider the advantage of a constant body temperature to be that enzymes work best at constant temperature. We should remember, however, that enzymes have a remarkable ability to adapt to different prevailing temperatures, for example through development of iso-enzymes, which are enzymes with the same function but with different temperature-dependent kinetics. The enzymic processes which sustain bacteria living in below-freezing sea water are basically the same as those which sustain bacteria in hot-water springs. Also, even in an animal which does have a constant body temperature, enzymes are required to act at different temperatures. The enzyme thrombin, for example, must play its role in preventing leakage from blood vessels whether the potential leak occurs in the human heart, at 37 °C, or a few seconds later in the finger tip, at less than 20 °C.

In those animals which do regulate their body core temperatures, many vertebrates and some invertebrates, the regulated temperature usually lies in the range of 32–42 °C. Having a body temperature above air temperature allows some metabolic heat to be dissipated without evaporating valuable body water (see McArthur and Clark, 1987). However, no one has advanced a valid physical, chemical or biological reason why the regulated temperatures lie in a specific range. A hypothesis, as good as any other, is that the range represents the ocean temperatures at the time at which terrestrial life emerged. Alternative explanations will be presented later.

Perhaps because it is relatively easy to measure temperature accurately when compared with many other ecophysiological variables, an enormous literature on thermal biology has accumulated over the past decades. We cannot hope to cover it extensively here and will therefore concentrate on key principles.

1.2 Terminology

1.2.1 Body temperature

Animal temperatures vary with the time of day and among and between species. In those animals which have a temperature different from environmental temperature, different parts of the body exhibit different temperatures. There is therefore no such thing as a single 'body temperature'. Representative temperatures usually are those of deep body tissue measured through a convenient orifice; standard practice is to use the rectum or cloaca.

Examples of animals which regulate temperature well are:

Monotremes	30–31 °C
Marsupials	35–36 °C
Eutherian mammals	36–40 °C
Birds, non-passerine	39–40 °C
Birds, passerine	40–41 °C

The amplitude of the circadian rhythm in body temperature is related to body size, with smaller animals tending to exhibit the greater amplitude, a consequence of their lower thermal inertia. There are exceptions; the camel has an unexpectedly large circadian variation. Another important rhythm of body temperature is that associated with reproductive cycles; in most women, for example, body temperature rises about 0.5 °C at ovulation.

Relatively few species (all mammals or birds) regulate body temperature accurately all the time. A great many more species regulate well for part of each day, by using thermoregulatory behaviour. The temperature which they then select to maintain is known as the 'preferred' or 'eccritic' body temperature for that species, though a better term is 'selected' body temperature.

1.2.2 Homeothermy vs. poikilothermy

Because the historical terms 'cold-blooded' and 'warm-blooded' are illogical, the terms 'homeothermic' and 'poikilothermic' were introduced to describe animals with a well-regulated and a variable body temperature respectively. Another classification of thermoregulation describes animals as 'endothermic' (producing heat within their own tissues to thermoregulate) or 'ectothermic' (relying on gained heat from the environment when thermoregulating).

The categorisations endothermic–ectothermic and homeothermic–poikilothermic are independent (Figs 1.2 and 1.3). In theory, any ectotherm can be homeothermic, given a suitable habitat. However, if an animal is habitat independent and homeothermic, it must be an endotherm. Not all animals display the same thermoregulatory pattern at all times. Some alter the pattern according to environmental circumstances. For example, some moths, beetles and carpenter bees, usually ectotherms, employ endothermic mechanisms before and during flight. Nor do all species in an order have the same pattern; most mammals are homeothermic endotherms, but the naked mole rats, fossorial mammals, appear to be typical poikilotherms.

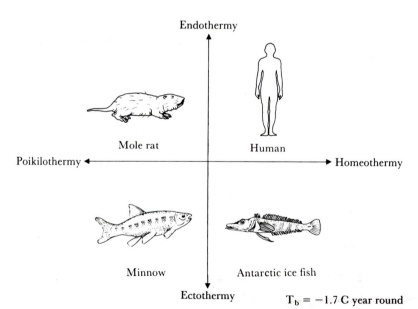

Endothermy

Poikilothermy ← → Homeothermy

Mole rat   Human

Minnow   Antarctic ice fish

Ectothermy $T_b = -1.7\,C$ year round

Figure 1.2 Categorisation of thermoregulatory abilities, with typical examples. There are very few true homeothermic ectotherms, though many ectotherms do maintain an almost constant body temperature, given an appropriate thermal environment.

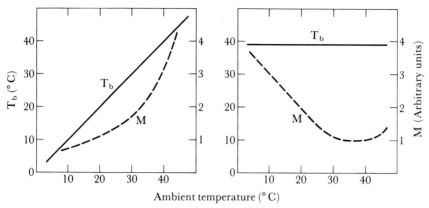

Figure 1.3 Body temperature (T_b) and metabolic rate (M) of a typical poikilothermic ectotherm (left) and homeothermic endotherm (right) in relation
to ambient temperature.

1.2.3 Behavioural vs. physiological thermoregulation

Caution is always advocated in discriminating between 'behavioural' and
'physiological' phenomena because they are indistinguishable at the level of
the physics and chemistry (physiology) of cells. Nevertheless, a strong con-
vention exists which describes the use of site selection, position and posture
for temperature regulation as behavioural thermoregulation, and the use of
endogenous metabolic heat production, cardiovascular adjustment, and
evaporation as physiological thermoregulation.

1.3 Physics of heat exchange

Heat is a synonym for the total kinetic energy of all the molecules in a system,
and all systems with temperatures above absolute zero ($-273\,°C$) contain
heat. Temperature, on the other hand, is a measurement of the mean kinetic
energy of the molecules in a system. When two systems with different temperatures are placed in contact, heat flows from the one with the higher
temperature to the one with the lower temperature. The rate at which heat
will flow between them cannot be determined from the difference in their
temperature alone (Bartholomew, 1982). Let us examine the most important
avenues of heat gain and heat loss which animals experience, and the most
important factors affecting the rate of heat exchange.

1.3.1 Conduction

Conduction is the movement of heat by interaction of adjacent molecules
without the mass motion of the medium through which the energy transfer
takes place.

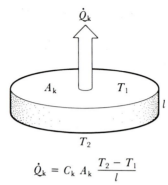

$$\dot{Q}_k = C_k \, A_k \, \frac{T_2 - T_1}{l}$$

Figure 1.4 Rate of heat flow ($\dot{Q}_k$) through a segment of conductor of thickness *l*
and area A_k, and face temperatures T_1 and T_2.

The rate of heat transfer by conduction (Q_k) through a segment of material can be expressed by the equation shown in Fig. 1.4. The thermal conductivity coefficient (k) is a number which describes the ease with which heat flows by conduction through given material. For example, the k value for silver is 0.41, for wood 0.0001 and for air 0.000 024 W mm^{-1} °C^{-1}. The last figure explains the important insulative function of air trapped in animal fur and protective clothing.

From the conduction equation we can deduce that the most important factors affecting heat gain or loss from an animal by conduction, other than the k value, are the surface areas involved and the temperature gradients between the animal and the environment.

Heat transfer by conduction therefore is of great importance to those animals in which there is a large contact area with a solid surface, such as snakes and many invertebrates. High rates of heat transfer with the solid substrate may have contributed to the development of poikilothermy in fossorial mammals like golden moles and mole rats. In humans walking or standing, even on snow or ice, very little heat is gained or lost by true conduction as only small areas of the body (soles of the feet) are in direct contact with a solid medium.

1.3.2 Convection

Consider a naked human with a skin temperature of 35 °C standing in a room with completely still air (air temperature 25 °C). Comparatively little heat will be lost through the soles of the feet because of the small surface area. However, because of the temperature gradient of 10 °C between the entire surface of the rest of the body and the surrounding air, kinetic energy will be transferred from the surface of the skin to the boundary layer of air adjacent to the skin. As the air molecules gain kinetic energy, they will move upward and away from the skin creating fluid currents. The heat loss is no longer by conduction, but by means of 'free convection'. If we were now to turn on a fan, heat loss will be increased even further by so-called 'forced convection', because the boundary layer is removed more rapidly by the external wind. Convection is thus the movement of heat through a fluid (either liquid or gas) by mass transport in currents.

For an animal in a moving fluid, the rate of heat loss is governed by the equation:

$$\dot{Q}_c = C_c A_c V^n (\bar{T}_s - T_f) \qquad [1.2]$$

where $\dot{Q}_c$ (W) is the rate of convective heat loss, A_c (m²) the area of the animal participating in the convective heat exchange, V (m s⁻¹) the fluid speed, $\bar{T}_s$ (°C) the mean surface temperature of the animal, and T_f (°C) the fluid temperature. The exponent n depends on the size of the animal and the range of fluid speeds over which the equation is applied: for a typical mammal or bird in air, $n \simeq 0.5$, so that convective heat transfer is approximately proportional to the square root of wind speed. The coefficient C_c varies not only with animal size but also with physical properties of the fluid, including density: convective heat loss rate reduces with altitude. More important, convective heat transfer is so rapid in water that it is very costly for any aquatic animal to maintain a body temperature different to water temperature, and most do not do so. Naked humans are unable to regulate body temperature in water colder than about 15 °C, a temperature above that of most of the world's ocean water. For a more detailed analysis of the physics of forced and free convection, see Mitchell (1974).

Remember that when air or water temperatures are above body temperature then heat gain is accelerated by convection. For this reason the Bedouin shelter from hot desert winds while naïve Europeans seek out wind on the mistaken understanding that wind always has a cooling effect. Wind speed is of crucial importance at low air temperatures, and the wind chill factor is a measure of the combined effect of air temperature and wind on cooling and evaporation. A person exposed to an air temperature of −40 °C and a 50 km h⁻¹ wind would lose heat as if exposed to −80 °C in still air. The record low for our planet is −87 °C (1958) at USSR Antarctic base. The incremental effect of wind speed is, however, particularly noticeable at low wind speeds for two reasons. One derives from the convective heat transfer equation: $\sqrt{V}$ increases three-fold between 0.1 and 1 m s⁻¹, but only 15% between 30 and 40 m s⁻¹. The other is biological: at fairly low speeds the natural lie of the animal's coat (pelage) is disturbed, thereby allowing cold air currents to reach the skin.

The fluid speed (V) in the convective heat transfer equation (eqn [1.2]) is the relative speed of body and fluid, so the equation refers as well to an animal moving through a fluid or to a fluid moving past an animal. Flying and swimming animals particularly are subjected to high convective heat loss; some birds will deliberately dangle their legs to enhance heat loss when flying in hot environments.

1.3.3 Radiation

Radiation is energy exchange by means of electromagnetic energy which travels at the speed of light and needs no medium of propagation. The rate of radiant heat transfer between an animal and its environment is given by

$$\dot{Q}_r = C_r A_r ((\bar{T}_s + 273)^4 - (\bar{T}_r + 273)^4) \qquad [1.3]$$

where $\dot{Q}_r$ (W) is the rate of radiant heat loss, A_r (m²) the area participating in the radiant heat transfer, and $\bar{T}_r$ (°C) the mean radiant temperature of the surroundings, that is, the surface temperature of objects exchanging radiation with the animal. The coefficient C_r incorporates the Stefan–Boltzmann constant as well as the emissivities of both the animal's surface and of the solid surfaces making up the radiant environment.

Heat from animal bodies is emitted in the middle infra-red (5–20 µm). Heat is gained by animals from direct solar radiation, mostly within the visible range (0.4–0.7 µm), but also from reflected visible energy as well as re-radiated long-wave radiation. In the case of direct solar radiation and reflected sunlight the colour of the animal has an important influence on the coefficient C_r as more energy within these wavelengths is reflected by light colours than by dark colours. Dark colours, in turn, absorb more energy than light colours under these conditions. It is important to note, however, that direct sunlight heats up the surface of the earth and this energy is re-radiated in the form of long-wave radiation, the absorption of which is not affected by colour. Based on these principles one would expect animals in polar regions to be black where the incident radiation is mostly visible energy (sunlight and reflected sunlight from snow and ice) and in hot deserts we would expect animals to be light-coloured. However, there is no simple relationship between thermal environment and animal colour. Indeed, some colours seem paradoxical. Polar bears are white and therefore one expects them not to be able to use solar warming well. In the white-hair coat (pelage), however, the individual hairs may act like optical fibres and reflect short-wave energy through the pelage to the surface of the skin. Also many desert animals are pitch-black in colour. The desert raven's black plumage absorbs a large amount of solar radiation and the tips of the feathers can reach temperatures as high as 80 °C. Very little of this heat actually reaches the skin surface, however, as the air trapped beneath the plumage acts as an insulative barrier between the feathers and the skin. Also, the tips of the feathers reach a temperature well in excess of the air temperature even on a hot day and the high thermal gradient encourages heat loss by both convection and radiation. Similar circumstances may prevail in the case of other black desert animals.

1.3.4 Change of state

Change of state in biological systems involves either the change of state of liquid water into water vapour or vice versa and either the uptake or release of the latent heat of vaporisation or condensation respectively. Rate of heat loss via this avenue is influenced mainly by vapour pressure gradients, surface areas and wind speed.

Consider a naked human standing in direct sunlight on a hot day. If the environmental temperature exceeds the body surface temperature, the only way in which the subject can lose heat is by evaporating water by sweating, as humans do not exhibit thermal panting. The lower the water vapour pressure of the air the faster the sweat will evaporate. If the wind speed increases, the boundary layer next to the skin, which is saturated with water vapour, will be disrupted, thereby increasing the vapour pressure gradient from the skin to the surrounding air and accelerating evaporative cooling. Naturally the larger the surface area involved, the greater the potential cooling effect. Per kilogram of body mass, humans lose more water through sweating than any other member of the animal kingdom, and access to such powerful cooling helps to maintain the temperature of their central nervous systems at a very constant temperature.

Because rates of evaporation and convection both depend fundamentally on the properties of the boundary layer near the body surface, the equation describing evaporative heat loss from the body surface looks very like the convection equation:

$$\dot{Q}_e = C_e A_e V^n (P_{H_2O_s} - P_{H_2O_a})$$ [1.4]

where $\dot{Q}_e$ (W) is the rate of evaporative heat loss, A_e (m^2) the area participating in evaporation, $P_{H_2O_s}$ (kPa) the saturated water vapour pressure at body surface temperature, and $P_{H_2O_a}$ the water vapour pressure of ambient air. The exponent n has exactly the same value as in the convection equation, that is, about 0.5 for typical mammals. The coefficient C_e depends on the same physical properties as C_c, but also includes a factor accounting for how wet the body surface is. Because this factor is very difficult to measure, the equation is seldom used. Rather, use is made of the rate of mass loss of the animal as an indirect measure of evaporation rate. Then:

$$\dot{Q}_e = \lambda \dot{m}$$ [1.5]

where λ (J g^{-1}) is the latent heat of evaporation of water, and $\dot{m}$ (g s^{-1}) the rate of mass loss.

If an animal evaporates water predominantly via the respiratory tract, the surface evaporation equation no longer applies, but the mass loss equation does. Evaporation from the respiratory tract can also be determined by measuring ventilation rate and the water content of inspired and expired air.

The driving force for evaporation is the water vapour pressure gradient.

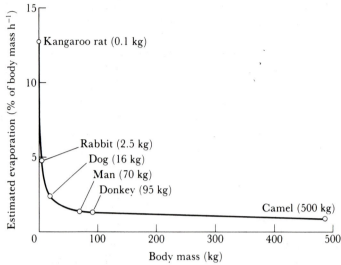

Figure 1.5   Estimated rates of evaporation of water necessary to maintain a constant body temperature for various mammals standing at rest in a typical desert environment. Heat load is proportional to body surface area, and available water is proportional to body mass, so small animals, with a high surface area relative to mass, are disadvantaged. Redrawn from Schmidt-Nielsen (1964).

Meteorologists like to express the water vapour concentration in air in terms of the relative humidity. Simplistic deductions based on relative humidity can, however, be misleading. For example, evaporation from a human takes place faster in air at 90% relative humidity, 15 °C temperature, than in air at 40%, 30 °C, because vapour pressure rises rapidly with temperature (see Chapter 2). It is perfectly possible to evaporate sweat into air with 100% relative humidity, provided the temperature is low enough.

The combination of high air temperature and high humidity, on the other hand, results in very high vapour pressure, and poor or absent evaporative cooling. Based on these physical principles, Richards (1973) speculates that when the unfortunate victims of the infamous 'Black Hole of Calcutta' incident were crammed into a small prison cell, air temperature must have risen rapidly to skin temperature. The prison cell was poorly ventilated and consequently the atmospheric air soon became saturated from the sweat and expired air of the prisoners. Body temperatures must have risen swiftly under these conditions with rapid onset of heat death. Similar conditions prevail inside space suits, and a good deal of the paraphernalia carried by astronauts on space walks is concerned with removing heat.

Birds and mammals are the only animals that regularly employ active evaporative cooling. Some lizard species will pant but usually as an emergency measure only. Birds do not sweat but most do employ thermal panting. Small mammals never sweat because of the danger of dehydration (see Fig. 1.5). For example, if a small mouse were to sweat it would become fatally dehydrated within a few hours. Instead, they use behavioural mechanisms to escape from heat to a more favourable microclimate. Again, in an emergency, small mammals like rats and bats will cover their body surfaces with saliva. Some large animals both sweat and pant (e.g. most large antelope, sheep and cattle), some only pant (e.g. dogs, pigs and wildebeest). Rhinos sweat but elephants do not; they dilate the blood vessels to their large ears, which then act as thermal windows for radiant and convective cooling. Flapping of the ears increases convective cooling through these thermal windows.

1.3.5 Newton's Law of Cooling

Newton's Law of Cooling, expressed in its simplest form, states that the rate of cooling of an inert body is proportional to the difference in temperature between the centre of that body and the surrounding medium:

$$\dot{T}_b = C_N(T_b - T_a) \qquad [1.6]$$

where C_N represents the Newtonian cooling constant of the particular body, T_b is the temperature of the body (and $\dot{T}_b$ its rate of change) and T_a is the ambient temperature. In other words, if the body is incapable of producing heat, the rate of heat loss will decrease exponentially as the body's temperature approaches that of the environment (see Fig. 1.6).

The concept provided by Newton's Law has been useful in analysing and testing various hypotheses in thermal biology. For example, if the heating

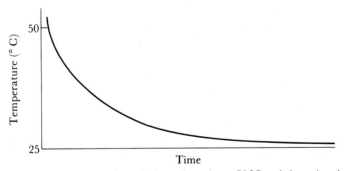

Figure 1.6 Temperature of a body heated to above 50 °C and then placed in an environment of 25 °C, cooling according to Newton's Law. The temperature (T_t) at a time t is given by the equation

$$T_t - T_f = (T_i - T_f)e^{-t/\tau}$$

where T_i and T_f are the initial and final temperatures, and τ the time constant.

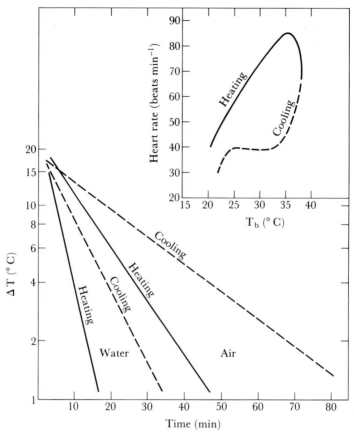

Figure 1.7 Difference in temperature (Δ*T*) between body core and environment, as a function of time, during heating and cooling of the Galapagos marine iguana (*Amblyrhynchus cristatus*) between 20 and 40 °C, in water and air. The rates of heating and cooling are much faster in water than in air, because of the much higher heat transfer coefficient at the animal surface. In both water and air, the rate of cooling is lower than the rate of heating, a deviation from Newton's Law, implying that the rate of heat transfer within the iguana differs during heating and cooling. Such differences usually result from a change in the cardiovascular status, which is reflected in the hysteresis in the relationship between heart rate and body temperature (*T*_b) during heating and subsequent cooling. Modified and redrawn from Bartholomew and Lasiewski (1965).

and cooling rates of the Galápagos marine iguana in air and water are drawn on a semi-logarithmic plot, a great deal of information can be gleaned from the results (Fig. 1.7). First, the rate of cooling and heating in water is swifter than in air because of the better heat transfer in water. Also, in both cases rates of heating are more rapid than cooling rates. For an inanimate body, rates of cooling and heating must be equal, according to Newton's Law, so unequal rates imply active thermoregulation by the animal. In the case of the marine iguana, it is adjustments in cardiovascular function, particularly vasomotor tone, which affect the transport of heat between the core and the periphery and vice versa.

Tracy (1972) has cautioned against the indiscriminate use of Newton's Law when studying the thermal adaptations of homeotherms. He argues that the effects of wind speed and animal size must be evaluated separately. Similarly, Bakken (1976) maintains that the analysis of cooling curves obtained by plotting log (body temperature − air temperature) vs. time is generally inaccurate.

1.3.6 Conductance

Notwithstanding its potential inaccuracies, Newton's Law has allowed the introduction of the concept of conductance into thermal ecophysiology, and measurements of conductance have allowed important insights into thermal adaptations. If the body undergoing Newtonian heating or cooling has mass m (g) and specific heat p ($J\,g^{-1}\,°C^{-1}$), then its rate of heat transfer with the environment $\dot{Q}$ (W) is

$$\dot{Q} = mp\dot{T_b} = mpC_N(T_b - T_a) \qquad [1.7]$$

The term mpC_N is called the conductance (K); it differs from the heat transfer coefficients described earlier in that it defines heat transfer from the core of the body, rather than its surface, to the environment, and is influenced by vasomotor tone, subcutaneous fat, etc.

If a non-evaporating animal reaches thermal equilibrium at a particular ambient temperature, then its heat loss to the environment must equal its metabolic heat production:

$$\dot{Q}_m = K(T_b - T_a) \qquad [1.8]$$

Hence, the slope of equilibrium metabolic heat production plotted against air temperature is the negative of the conductance.

1.3.7 Size, shape, time constants and thermoregulation

An animal's size has a profound effect on its thermoregulation. The most important size parameter affecting thermoregulation is the ratio of surface area to body mass. Bearing in mind the allometric relationship between the relative surface area and the body mass of an animal, which scales to the power of 0.66, the smaller an animal the greater the relative surface area and, all other factors being equal, the faster it will gain or lose heat. A desert ant crawling across a hot desert dune has an extremely high relative surface area and gains heat very rapidly from the environment. Nevertheless, if it can retreat to a cool micro-environment (shade of a pebble) it can unload its heat very rapidly and continue on its journey for a limited time. In contrast, an elephant has a huge volume and a small relative surface area. It gains and loses heat very slowly; it has high thermal inertia.

For similar reasons, juveniles of a sand-dune lizard can use the dying rays of the sun to remain active in the late afternoon, but have to bury in the sand in the middle of the day, when the adults are active (Seely *et al.*, 1990).

Small animals also tend to have high conductance, which exacerbates their rate of heat loss in cold environments. Small endotherms, depending upon environmental circumstances, will generally be obliged to produce more heat per gram of tissue than large animals, whereas large animals may experience difficulty in getting rid of excess heat even though they generate far less heat than small animals per unit body mass.

Animal shape also influences heat exchange. Having body elements with a filament-like geometry minimises radiant heat gain and maximises convective heat loss for animals exposed to the sun, so the ant's shape also contributes to its ability to be active in the desert sun. Similarly, the typically rotund shape of polar animals, like the polar bear or Arctic ptarmigan, maximises radiant heat gain and minimises convective heat loss.

An animal's size, shape and conductance all influence the rate at which it will reach a new equilibrium of body temperature after a change of thermal environment. The approach to a new equilibrium temperature is exponential, so the rate of approach can be expressed in terms of the 'time constant', the time taken for $(1 - 1/e) \approx 63\%$ of the full change in temperature to occur. The time constant may be measured as the negative reciprocal of a log-linear plot of body temperature against time, following a change in thermal status; changes in slope of this plot indicate active physiological thermoregulation.

The time constant is a useful parameter for interspecific comparisons of rates of heat transfer. It also quantifies well the responses of a single species to different thermal challenges. For example, the time constant for naked humans transferred from a neutral to a hot environment is about 40 min, while it is over 3 hours following transfer from a neutral to a cold environment, reflecting activation of heat conservation mechanisms.

1.4 Measuring temperature and heat transfer

In the past several decades the measurement of temperature of both animals and their environments has become progressively more accurate and easier to accomplish. The introduction of thermistors, electronically compensated thermocouples, thermopiles, infra-red thermometers and radiotelemetry have all contributed towards this improvement. For further details, and particularly to avoid common errors, the reader is referred to an excellent practical guide by D. M. Unwin (1980) called *Microclimate Measurements for Ecologists*.

Although standard micrometeorological measurements are of great value to the thermal biologist, it would be very useful to have a single index which expresses heat transfer. Several recent studies in the field have used the 'standard operative temperature'. Standard operative temperature is a concept originally developed by Gagge (1940) for analysing human heat transfer and is the temperature of an ideal isothermal black body enclosure with standard convection conditions, which would produce the same net sensible heat flow for the same animal surface temperature. It is therefore an index of sensible heat flow to and from the environment. It can be calculated from standard micrometeorological variables, but in practice it is measured by constructing a model of exactly the same dimensions as a representative animal of the species being studied. The model is fitted with thermometers and is covered with the pelt of the animal being studied. These taxidermic models can then be placed in strategic positions, for example in full sunlight, in burrows, in the shade or various positions above ground or in a tree while their temperatures are being monitored. Using this method Chappell and Bartholomew (1981) studied the thermal energetics of the antelope ground squirrel in its southern Californian desert habitat. They found that standard operative temperature in unshaded areas was 30 °C above the upper critical temperature of the ground squirrels and 20 °C above air temperature for much of the day. In these conditions the maximum time that the animals could tolerate surface exposure was 7–9 min. Their findings supported the hypothesis that thermoregulation in this ground squirrel involves cycles of transient hyperthermia followed by periods of rest in underground burrows when heat is dissipated passively. For further details of the method consult Bakken (1980) and for interesting application to measuring the energetic cost of free existence see Weathers and Nagy (1980).

Because the construction of taxidermic models for measuring operative temperature is quite involved and time-consuming and because the measurements they give are species specific, many thermal biologists choose to use the simpler, but similar, measure of 'black bulb temperature'. The standard black bulb or globe thermometer in meteorology is a hollow copper sphere 150 mm in diameter, painted matt black and fitted with a thermometer which measures the temperature at the centre of the sphere. The black bulb or globe temperature will integrate several environmental variables namely radiation, wind speed and air temperature. As long as an animal does not change emissivity, globe temperature is a reasonable index of the animal's heat transfer in the prevailing environmental conditions, though it clearly cannot account for evaporative or metabolic processes. Seely et al. (1990) have shown the advantage of globe temperature over air temperature and substrate temperature in explaining the way lizards employ the thermal mosaic on an apparently homogeneous desert sand dune.

An approach which combines aspects of the black bulb thermometer and the taxidermic model is to construct a metal cylinder the same size as the study animal, paint it black and fit it with a thermometer in order to monitor its interior temperature. Very small cylinders the size of ants or bees can be made. In this way Louw and Nicolson (1983) were able to measure a critical minimum temperature required for flying in carpenter bees with better predictive value than normal air temperature.

1.5 The heat balance and the complexity of heat exchange in an outdoor environment

Thermal flux through an organism is usually described by the deceivingly simple equation:

$$\dot{Q}_m + \dot{Q}_w + \dot{Q}_k + \dot{Q}_r + \dot{Q}_c + \dot{Q}_e = \dot{Q}_s \qquad [1.9]$$

where $\dot{Q}_m$ is the metabolic rate, $\dot{Q}_w$ is the rate of energy expenditure on physical work, $\dot{Q}_k$, $\dot{Q}_r$, $\dot{Q}_c$ and $\dot{Q}_e$ are the rates of heat transfer by conduction, radiation, convection and evaporation with the environment, and $\dot{Q}_s$ is the rate of heat storage, which equals zero when the organism is in thermal equilibrium. Although the concept is simple, the measurement of all the variables in an outdoor environment is exceedingly complex (see Porter and Gates, 1969, see Fig. 1.8).

Because of their complexity, complete heat balances have seldom been attempted. A famous study is the analysis of the heat balance of a lizard by Bartlett and Gates (1967). Studies on African mammals have shown that one of the most important avenues of heat gain in a hot arid environment is via re-radiation of long-wave radiation from the surface of the soil. These and other studies have also highlighted the importance of surface areas, particularly the profile area exposed to direct solar radiation, as well as the importance of thermal gradients between the organism and the environment. Also of importance is wind speed, the nature of the integument (pelage) and, in the case of man, clothing. These structures can act as thermal shields resulting in the exchange of heat at the surface of the animal without eliciting any physiological responses.

1.6 Control systems

When the first vertebrates emerged on land some 400 million years ago they immediately enjoyed the great advantage of extracting oxygen from a medium (air) with a high oxygen content and a low density. This advantage eventually led to the evolution of homeothermic, or endothermic, mammals and birds, with a constant body temperature and a metabolic rate some four to eight times greater than their ectothermic ancestors. The advantages of terrestrial life and physiological control over thermoregulation are many and varied but there is a price to pay, namely dehydration as a result of the ventilation of the respiratory surfaces. We will examine later the interesting adaptations that have evolved to compensate for the risk of dehydration. Let

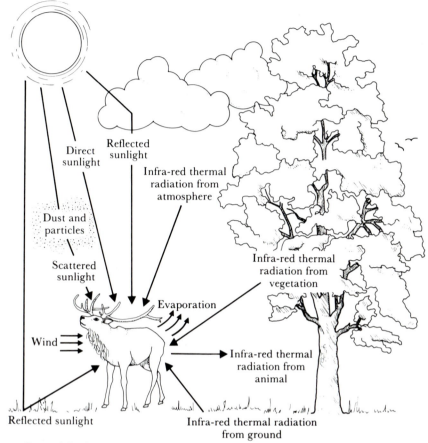

Figure 1.8 Avenues of heat transfer between an animal and an outdoor environment. Redrawn from Porter and Gates (1969).

us first examine the control systems involved in both endothermic and ectothermic thermoregulation. Temperature control lends itself to systems analysis, and there have been very sophisticated analyses of the control, particularly for humans.

In Fig. 1.9 a simple model depicts the central nervous system as a black box receiving afferent information from cold and warm sensors situated in both the periphery and the core of the body. This information is integrated and produces the appropriate efferent response. Sweating, shivering and non-shivering thermogenesis, and to a large extent panting, are restricted to true endotherms (mammals and birds), while ectothermic animals (the rest of the animal kingdom) can at best use only changes in vasomotor tone and behaviour. The latter two responses are naturally used extensively by endotherms as well.

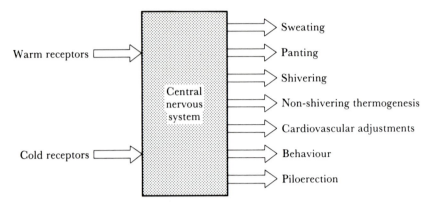

Figure 1.9 The thermoregulatory system as a black box which receives information from warm and cold receptors, and actuates appropriate effector mechanisms to dissipate, generate or conserve heat.

This simple model has been expanded frequently to provide mathematical and engineering models, expressed in the analogous terms of a proportional controller. In parallel with efforts to understand the systems characteristics of temperature control better, there has been a concerted effort to identify the anatomical elements and physiological function of the controller. Neurochemical and neurophysiological approaches were employed. In the neurochemical approach, putative neurotransmitters and their antagonists have been introduced into the nervous system. It has been possible to produce the entire repertoire of thermoregulatory responses in several animal species by micro-injecting minute amounts of various neurotransmitters at appropriate brain locations. The same neurotransmitter will elicit appropriate responses in species as diverse as the sheep and the giant monitor lizard, suggesting a very ancient origin for the neurochemical mechanisms. However, no unifying neurochemical model has emerged capable of explaining responses across species, nor in different thermal circumstances, and there are so many methodological problems associated with current techniques that one cannot be optimistic about achieving a coherent model.

The neurophysiological approach has been much more successful. Neurones capable of functioning as temperature sensors have been identified in peripheral tissue (especially the face and scrotum), and in the central nervous system, of several species (see Fig. 1.10). They have firing rates which either increase (warm sensors) or decrease (cold sensors) with temperature change in the physiological range, and their characteristics are very similar in all species studied, including ectotherms. The modern techniques of brain slice recording and neuronal culture have established that single neurones can act as biological thermometers without any connection to other neurones; on the other hand, there are neurones in the brain which function as interneurones capable not only of integrating afferent input from tempera-

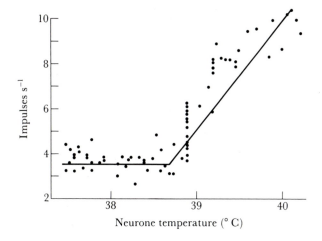

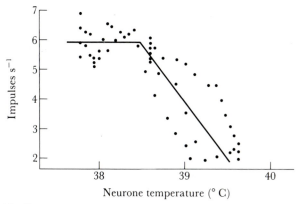

Figure 1.10   Temperature–activity patterns of hypothalamic neurones in a conscious rabbit. A miniature heat exchanger was implanted in the rabbit's hypothalamamus, and microelectrodes were used to measure firing rates of nearby neurones. Most hypothalamic neurones have a firing rate which is not affected by local temperature. Some have a firing rate which increases with temperature within the normal range of brain temperatures (top) or decreases with temperature over the same range (bottom). Redrawn from Hellon (1967).

ture sensors in different body regions, but also capable of integrating the thermoregulatory control system with other control systems, like that responsible for maintaining body fluid volume.

1.7 Metabolic rate and thermoneutral zones

Basal metabolic rate is a medical term invented for humans. It describes the minimum amount of energy required to support essential life processes such

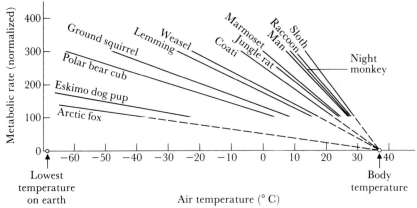

Figure 1.11   Regression lines fitted to measurements of the metabolic rate of
various mammals in relation to air temperature. The resting metabolic
rate of each species, in the absence of cold stress, is given the value
100%, making it possible to compare widely differing species. The
slope of the regression line is a measure of conductance, and Arctic
animals typically have much lower conductances than tropical animals.
Redrawn from Scholander *et al.* (1950a).

as heart function, respiration, muscle tonus, etc. when the subject is at
complete rest, in the post-absorptive state (i.e. food has not been eaten
during the past 12 hours) and in a thermally neutral environment. It is
seldom possible to comply with all these conditions when studying animals
and instead we describe the conditions of the experiment and rather speak of
standard or resting metabolic rate. If we expose an endothermic animal to a
wide range of temperatures and simultaneously measure its metabolic rate
we shall find a temperature range within which metabolic rate will not
change. The width of this so-called thermoneutral zone will depend largely
on the conductance and body size of the animal concerned. This important
principle is well illustrated in Fig. 1.11 which shows that well-insulated
Arctic animals, such as the Arctic fox, have very wide thermoneutral zones
and only raise their metabolic rates when exposed to temperatures as low as
−40 °C. In contrast, naked humans have a very narrow neutral zone and in
the case of some small mammals the zone is merely a single point on the
temperature axis.

If we recall the conductance equation (eqn [1.7]), we can see that polar
animals, as expected, have much lower conductances than tropical animals;
their slopes are much flatter. Also, the equation predicts that if the lines are
extrapolated to the *x*-axis they intercept this axis at body core temperature,
about 37 °C for most mammals. Interestingly, the extrapolated conductance
curve for birds does not intercept the *x*-axis at body temperature, presumably

because birds can change their shape and therefore their conductance markedly by fluffing their feathers out and changing their posture.

Because an increased metabolic rate represents an energy cost to the animal, various physiological adaptations have evolved to reduce this potential energy wastage. These include summer aestivation in the case of desert animals, hibernation during winter and even a daily torpor in the case of very small endotherms such as humming-birds and shrews.

The increase in metabolic rate with declining ambient temperature, shown in Fig. 1.11, is largely the consequence of shivering of skeletal muscle. Shivering is an emergency procedure and animals will go to great behavioural lengths to avoid having to employ it. Shivering produces a large amount of heat but not all the heat is stored, because the muscle activity induces increased blood flow, and the extra flow results in heat loss from the body surface. Human shivering efficiency (i.e. percentage heat stored) is about 50%, which is nevertheless much higher than the efficiency of exercise as a means of generating heat to overcome cold, about 20%. During voluntary activity of a muscle, shivering is inhibited, which means that the effects of exercise and shivering are not additive. Also, the pathways responsible for activating shivering in voluntary muscle are separate to the pathways serving motor function in the same muscles and can be traced back through separate spinal tracks to the posterior hypothalamus.

In addition to shivering, certain mammals, particularly small ones, are capable of producing large amounts of heat by so-called non-shivering thermogenesis (NST). To test if an animal is capable of NST it is first given the opportunity to become cold-adapted by exposing it to low ambient temperatures and preferably a declining photoperiod. It is then given noradrenaline while its heat production is measured either directly in a calorimeter or indirectly by recording the oxygen consumption. Animals capable of NST will respond with a rapid, almost instantaneous, increase in heat production. The mechanism whereby NST operates at the cellular level involves sympathetic stimulation of β-adrenoreceptors in brown adipose tissue (BAT), and the dissipation of proton gradients across the mitochondrial membranes in this tissue. We shall discuss the phenomenon in some detail in Chapter 3. Non-shivering thermogenesis is an efficient means of heat production and is extensively used by small mammals when emerging from torpor and hibernation, as well as by the neonates of large mammals, including man.

1.8 Colour

The colours displayed by vertebrate animals have been the subject of much speculation among zoologists, geneticists and animal scientists. The subject is a complex one, particularly when considering the effect of opposing natural selection pressures. For example, it may be beneficial for an animal to be white in colour to match its background (so-called crypsis) and thereby

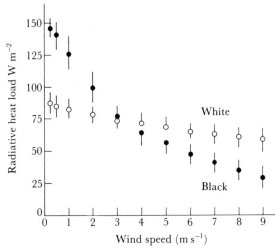

Figure 1.12   Radiant heat load transferred to the skin under black or white pigeon plumages, as a function of wind speed. Plumages were maximally erected, and exposed to $900\,W\,m^{-2}$ of simulated solar radiation. Redrawn from Walsberg *et al.* (1978).

escape predation, while at the same time black may be beneficial for thermoregulation and a brilliant display of colours may make it easier to attract mates. Under conditions of natural selection these opposing selection pressures are resolved by the most important vector in the complex of selection pressures. Study of animal colour proceeds from a misleadingly anthropocentric base too. Many non-human animals are completely colour-blind and those with colour vision may have completely different spectral sensitivity.

In addition to these complexities, we must also realise that the plumage and pelage of birds and mammals are not merely plain physical surfaces to which we can assign an absorption or reflectance value. Of great importance to the thermal biologist is the degree to which heat will penetrate through the pelages or plumages of different colours and thereby affect the heat gain of the animal. The biologist must also study heat exchange under natural conditions, when factors other than radiation, such as wind speed, become important considerations when evaluating colour. For example, Walsberg *et al.* (1978) have shown that, although black pigeons absorb a greater radiative heat load than do white pigeons in still air, this difference is progressively reversed as the wind velocity exceeds $4\,m\,s^{-1}$. Similar results were obtained by Lombard (1989) working on the plumages of black cormorants and white gannets in a wind tunnel. Walsberg (1983a) believes that irradiation penetrates more deeply into white plumages and heat gain is therefore less affected by the surface disruption of the plumage at higher wind speeds (Fig. 1.12).

Plate 1.1   Polar bears are protected from the cold by a thick pelage and layer of subcutaneous fat. It is thought that the white hairs act as optic fibres to convey short-wave radiation from the surface of the pelage to the skin surface, thereby warming the skin. (Photo: courtesy of Tom Ulrich.)

Surface reflectance of a pelage can therefore provide misleading data and penetrance of the radiation should always be measured. Øritsland and Ronald (1978) found that the reflectance values for three dark brown *Mustela* species were very similar (0.30–0.35), whereas the transmission of energy through their pelages varied two-fold. Also, Grojean *et al.* (1980) maintain that hollow, transparent hairs in polar bears act like optic fibres to guide short-wave radiation towards the darkly pigmented skin, thereby increasing the depth of penetration of the radiation and consequent heat gain. These results, in conjunction with the excellent adaptation exhibited by pitch-black ravens in hot deserts, suggest the almost absurd idea that black may be the ideal colour for thermoregulation in the desert and white ideal for the polar regions.

Remember also that colour affects only the reflectance and absorption of short-wave radiation (mostly within the visible range) and that the absorption of long-wave radiation is not affected by colour (see Fig. 1.8). In the desert large amounts of short-wave radiation are absorbed by the desert surface, which then heats up and re-radiates this energy as long-wave radiation. It follows that small animals living very close to the substrate will gain heat largely through long-wave re-radiation and colour becomes less important. In polar regions the high reflectance of snow and ice causes high levels of short-wave radiation (including ultraviolet radiation), hence the fairly rapid tanning of skiers even under very cold conditions. Evaluating colour is therefore a complex procedure and we must remember that many animals, by merely altering their orientation, posture or seeking out a favourable microclimate can readily compensate for the relatively small thermal disadvantage due to their colour. Nevertheless, it is an important ecophysiological concept as the following simple examples should show.

1.8.1 Chameleons

It is generally accepted today that the dramatic colour changes which chameleons exhibit serve largely to facilitate behavioural thermoregulation. Chameleons are ectothermic animals and rely on sunbasking and shade-seeking to regulate their body temperatures. When they are cold they disperse the melanin pigment in their integumentary cells so as to facilitate the absorption of short-wave, visible radiation. The colour changes are accompanied by changes in the shape of the body and the orientation of its long axis. For example, when warming up chameleons take on the shape of a broad leaf and orientate laterally towards the sun's rays. When they are exposed to excessively high temperatures the long axis of the body is orientated parallel to the solar beam, thereby reducing the profile area exposed to direct radiation, and they turn a pale colour (Burrage, 1973).

1.8.2 Amphibians

Generally speaking, amphibians are not efficient thermoregulators and remain in shaded moist habitats. Nevertheless, there are several interesting exceptions. For example, we have found that certain species of tree frog are able to blanch completely white when exposed to strong solar radiation. In cool conditions, when they need to raise their body temperature the frogs release the pituitary hormone melanophore-stimulating hormone (MSH), which disperses the melanin granules and they become almost black, so maximising the absorption of visible radiation. These little frogs are ecophysiologically interesting in other ways too; they are almost waterproof, which is most unusual for an amphibian (Withers et al., 1982; Kobelt and Linsenmair, 1986).

1.8.3 Birds

The colours of birds range across the complete spectrum from drab grey through pure white, pitch-black to the brilliance of humming-birds. Colour in birds is employed for a variety of reasons: crypsis, thermoregulation and optical signalling are the most important. Optical signalling and crypsis have obvious advantages but the thermoregulatory importance of colour is more subtle. For example, the Cape cormorant, a pitch-black bird, tends to over-heat when nesting on rocky islands. To cool itself it employs a form of rapid thermal panting. Because these birds have no access to fresh water one would expect that thermal panting would cause excessive concentration of body fluids, but apparently the efficient salt glands that discharge in their nasal sinuses compensate adequately for this stress. Why then are these animals black, rather than white like the gannets which nest next to them? The answer is to be found when one observes them foraging. They dive deeply in chase of their favourite prey, the pilchard, and to reduce buoyancy their feathers are hydrophilic or 'wettable'. This property brings the cold sea water ($12\,°C$) into direct contact with the surface of their skins ($35\,°C$) and heat drains rapidly from their bodies. As soon as foraging is complete, however, they return rapidly to the shore, spread their wings and orientate the maximum surface area to the sun. It is then that the black colour becomes a major advantage; they soak up solar radiation and warm their bodies without wasting valuable nutrients by shivering. The white gannets, on the other hand, have hydrophobic feathers and do not become wet during their short power dives after their prey, which is closer to the sea surface. They therefore have no need to enhance their sun-basking and their white colour with its high reflectance is an advantage during nesting. Analysis of the advantage of the cormorant's black colouring shows that interpretation of why an animal is a particular colour requires a knowledge of the animal's total ecology and life style. This point is further illustrated by the bedouin goat example described below, and the desert raven mentioned previously.

1.8.4 Mammals

The effect of natural selection on the colour of mammals is equally complex. The colour patterns of wild ungulates in the savannahs of Africa are remark-ably similar. With obvious exceptions, the pelage is usually a fawn to light fawn in colour. This colour is probably a compromise or optimisation to ensure reasonable reflectance of incident solar radiation as well as good crypsis against the blonde to light brown colour exhibited by the savannah vegetation for most of the year. The springbok (Fig. 1.13) has been studied in considerable detail (Hofmeyr and Louw, 1987). On cold mornings it orien-tates the long axis of the body laterally to the sun's rays and the dark brown stripe that runs bilaterally along the flanks facilitates the absorption of solar heat. As air temperatures begin to rise, the animals start to orientate parallel to the solar beam, thereby reducing the profile area exposed to direct solar radiation by as much as 50% (Fig. 1.14). The white colour of the rump and face also increases reflectance when this orientation is employed. Finally, the ventral surface in the springbok, as in many antelope, is pure white, which assists in reflecting short-wave radiation that is reflected off the substrate.

In the context of the ecophysiology of African mammals the black and

Figure 1.13   Lateral view of a springbok, illustrating its colour pattern: fawn shading, dark brown (diagonal lines) and white (remainder).

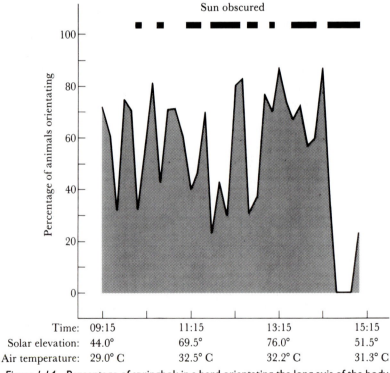

Figure 1.14   Percentage of springbok in a herd orientating the long axis of the body towards the sun. Bars at the top indicate when the sun was fully or partially obscured.

white stripes of zebras still remain a mystery. Some zoologists believe that the stripes enhance crypsis but the evidence is not convincing. We have measured the surface temperature of the stripes and, as expected, the skin surface covered by black stripes heats up more rapidly, and for short periods black stripes can attain a higher equilibrium temperature than the white stripes can. Zebras employ the same orientation behaviour as springbok, described above, and it is noteworthy that the ratio of black stripes to white is much higher on the flanks than on the rump. This pattern would facilitate warming when the animal is orientated at right angles to the solar beam and increase reflectance when the rump is presented to the sun. A really convincing explanation of the zebra's remarkable colour pattern is, however, still awaited.

Any consideration of the colour of ungulates would be incomplete without discussion of that fascinating small ruminant, the bedouin goat. Amiram Shkolnik of Tel-Aviv University has shown clearly that, in winter conditions, when the nutritional stress is at its peak in the Negev Desert, the metabolic rate of black goats standing in the sun is 25% lower than that of white goats. This difference apparently arises because the black goats absorb more solar radiation, and so have to expend less energy in shivering to maintain their body temperatures. Over a long dry winter the difference in metabolic rate can mean the difference between survival and death and the Bedouin have for centuries favoured the black colour in their selective breeding practices. The black colour of the goats is a disadvantage in the hot desert summer and black goats lose more water by sweating and panting than the white goats. Fortunately, however, bedouin goats have a remarkable ability to store very large amounts of water in their rumens and, even in summer conditions, only need to drink every second day.

1.9 Evolution of endothermy

Endothermy and the ability to maintain a constant body temperature would seem to have obvious advantages. Endothermic homeotherms, *inter alia*, maintain the temperature of their nervous systems at a constant level, locomote rapidly, have invaded cool and cold regions of the globe, and have become nocturnal predators. All these attributes have advantages, but care should be taken to avoid the anthropocentric view that endothermy is the apogee of evolutionary development. One need only mention the overwhelming success of the insects or the very extensive adaptive radiation of the teleosts to challenge that view. Moreover, endothermy has a high energy cost, requiring a metabolic rate 4–8 times greater, per gram of tissue, than that of most ectotherms. The energy demand, in turn, requires an increase in the respiratory ventilation rate and consequently increases the danger of both desiccation and disturbance to acid–base balance. Food requirements are also increased. It would be more circumspect to conclude

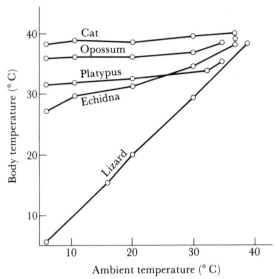

Figure 1.15 Variations of body temperature in a reptile (lizard), two monotremes (echidna and platypus), a marsupial (opossum) and a placental mammal (cat) caused by exposure for 2 hours to ambient temperatures between 5 and 40 °C. Redrawn from Martin (1980).

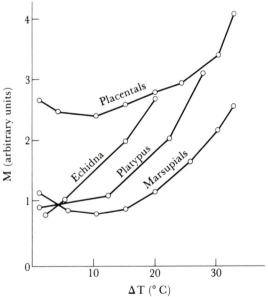

Figure 1.16 Metabolic rates of typical placental mammals, typical marsupials, and two monotremes (echidna and platypus) plotted against the difference between body temperature and ambient temperature, (ΔT). The curves of placentals and marsupials represent the results of experiments on several species of each kind. Redrawn from Martin (1980).

that endothermy, in spite of representing a quantum jump in evolutionary history, has conferred major advantages only to certain classes of animals within their specific niches. Humans have naturally exploited their advantages to the fullest through their concurrent intellectual development, which has allowed them to gain ascendancy over all other animals.

Much has also been made of the differences in body temperature among monotremes, marsupials and eutherian mammals as well as in their responses to cold (Figs 1.15 and 1.16). The initial inference that was drawn suggested that thermoregulation in the non-eutherian mammals was primitive and still evolving towards that of the eutherians. Today, however, these differences are considered to be the result of specialisation, analogous to that of the oryx antelope, which apparently can survive body temperatures of 45 °C in the desert. We also now know that some eutherian mammals, like naked mole rats, are poikilothermic, apparently more 'primitive' than the echidna.

The high metabolic rate of endotherms is achieved largely through a greater concentration of mitochondria and apparently also through the occurrence of two different functional forms of mitochondria in endotherms. For example, when comparing ectothermic mitochondria (frog) with endothermic mitochondria (rat) Akhmerov (1986) found that he could distinguish an endothermic mitochondrion in which oxidative phosphorylation was uncoupled, thereby producing heat and not ATP. The respiration of succinate in the absence of ADP was also five times greater in rat mitochondria than in frog mitochondria. The greater concentration of mitochondria in endotherms imparts additional advantages to these animals, not least of which is the ability of endotherms to undertake vigorous sustained exercise, because of their ability to maintain high rates of aerobic metabolism. This phenomenon has been examined in some detail by Bennett and Ruben (1979) and Bennett (1987), who conclude that high and stable body temperatures improve locomotor performance (Fig. 1.17). However, Bennett and Ruben (1979) caution that ectothermic vertebrates should not necessarily be considered easy prey for endotherms. The ectotherms' lower energy requirements for thermoregulation and efficiency of food utilisation allow them to occupy niches entirely unsuitable for endotherms. Their escape behaviour is often very successful and they can even pursue endotherms, provided that they need only short bursts of activity.

To summarise, then, endotherms and ectotherms have many thermoregulatory attributes in common, including similar control circuitry. Endotherms do, however, have special attributes, which must have evolved over a long period of geological time. There were probably intermediate forms maintaining lower body temperatures, such as the 25 °C of modern tenrecs. Once endothermy had been established, various 'spin-off' advantages could develop such as improved and continuous cortical function, improved care of young, and lactation. Improved cortical function may have led to increased

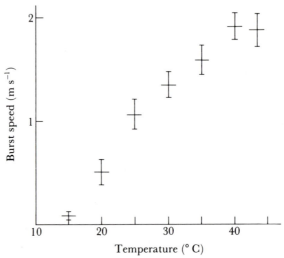

Figure 1.17 Burst speed as a function of body temperature in the desert iguana *Dipsosaurus dorsalis*, shown as mean ± 95% confidence interval. Data from Bennett (1980).

relative brain size or encephalisation. The possibilities for speculation are endless and fascinating but one should guard against simplistic explanations of such a complex process of evolution.

1.10 Fever (pyrexia)—aberration or adaptive response?

Almost everyone has experienced the discomfort which a fever causes—the cold shivers and eventual profuse sweating are well-known symptoms of this condition. In certain diseases such as malaria these responses are greatly exaggerated and very debilitating for the patient. What may be just a temporary unpleasant experience for a human may be lethal for other animals, where debility may mean inability to forage, or to escape a predator. The question therefore arises whether fever is merely a symptom of the disease or does it have survival or adaptive value? Before answering this question let us examine how thermoregulation changes during a bout of fever, as well as the physiological mechanisms involved in the production of fever.

During the first phase of fever a human patient feels cold and may experience cold shivers. This usually prompts him to clothe himself more warmly or use warmer bedclothing. The shivering will eventually disappear as the body temperature reaches a new elevated set point. Sweating then often occurs when the fever resolves. During the fever, the human mammal uses both behavioural and physiological responses to maintain body temperature at a new level. It is important to note that fever is not an uncontrolled hyperthermia but constitutes precise thermoregulation about a new elevated

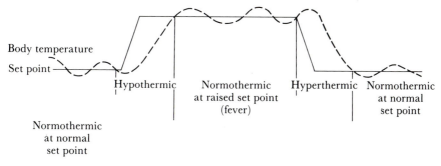

Body temperature

Set point

Hypothermic   Normothermic   Hyperthermic   Normothermic
             at raised set point   at normal
             (fever)   set point

Normothermic
at normal
set point

Figure 1.18   The thermal course of a fever. The system behaves as if the set point for body temperature were elevated at the beginning of the fever, and returned to normal at the end of the fever. Thermoregulatory effector mechanisms are initiated to bring the actual body temperature to the set point, as if the patient is hypothermic at the beginning of the fever and hyperthermic at the end. Redrawn from Bligh (1973).

set point (Fig. 1.18). A process similar to that in humans occurs in many other species of endotherm too.

The sequence of physiological events which accompany fever is complex and not yet completely understood. The flow diagram in Fig. 1.19 is a broad outline of these events. A variety of stimuli such as antigens, chemicals and particularly microbial agents can entrain the fever response. Exposure of an animal to these stimuli leads to the production of an endogenous pyrogen by the white blood cells (mostly monocytes and macrophages). Several endogenous pyrogens have been identified and are now even cloned. The endogenous pyrogen induces several secondary processes, one of which is a disturbance to the neurochemistry of the brain, which results in the elevated body temperature. Other processes include release of proteins by the liver, changes in serum metal ion concentrations, and a variety of hormonal events.

We return now to the question of whether or not fever has survival value. There is little doubt now that the release of endogenous pyrogens has a beneficial effect on the host. For example, they reduce the amount of circulating iron in the bloodstream, thereby inhibiting the reproduction of invasive pathogens.

The subsidiary question of whether the pyrexia, the elevated body temperature, contributes to the survival value remains unresolved. Hippocrates (Ancient Greek Father of Medicine) thought fever was advantageous, but modern doctors tend to treat fevers with antipyretic drugs, such as aspirin, which prevent the pyrexia without blocking other components of the fever. The first modern attempt to answer this question was made by the American physiologists Linda Vaughn, Harry Bernheim and Matt Kluger, who realised that lizards, because they are ectotherms and good behavioural thermoregulators, would be ideal material with which to test the hypothesis; no one had previously envisaged that ectotherms might develop fever. They first

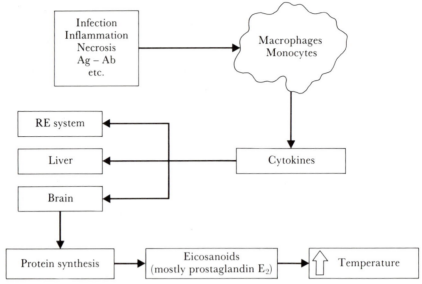

Figure 1.19 Outline of the biochemical events resulting in fever. In response to infection, inflammation, necrosis, antigen–antibody reactions and other stimuli, macrophages and monocytes release cytokines (inter-leukin 1, interleukin 6, tumour necrosis factor) which act on many organs, including the brain, where they initiate further protein synthesis and release of eicosanoids (primarily prostaglandin E$_2$). Modified from Mitchell and Laburn (1985).

placed desert lizards (*Dipsosaurus dorsalis*) in a thermal gradient which allowed them to select body temperature behaviourally, normally about 37 °C. When these lizards were injected with a bacterium pathogenic to reptiles (*Aeromonas hydrophila*), they selected a higher body temperature, about 42 °C, by remaining for longer periods at the warm end of the gradient (Vaughn *et al.*, 1974). Subsequently, Kluger *et al.* (1975) performed an even more significant experiment by again injecting *D. dorsalis* with live *A. hydrophila* and dividing them into five groups, which were held in incubators at 34, 36, 38, 40 and 42 °C, respectively. The lizards were unable to thermoregulate behaviourally in the incubators and merely acquired the temperature of the surrounding air. The results of the experiment showed a significantly higher rate of survival at higher temperatures, which suggests that fever may indeed have survival value.

More recently, Boorstein and Ewald (1987) carried out an even more sophisticated experiment using the grasshopper *Melanoplus sanguinipes*. They allowed the grasshoppers to eat food contaminated with a natural pathogen. The infected grasshoppers immediately sought out a warmer environment, and achieved an elevated body temperature. If prevented from attaining the high body temperature, the grasshoppers had severely compromised survival; again, pyrexia seemed to have survival value.

A survival value for fever would please the Darwinist who believes that all life processes must at one time or another have had a selective advantage to allow their development. If fever is as universally operative as the above studies would suggest, then it may have very ancient evolutionary origins; it has even recently been observed in a species of leech (Cabanac, 1989). However, as any biologist might expect, as further studies on the potential survival value of fever have accumulated, the picture has become increasingly complex and ambiguous (see Mitchell *et al.*, 1990). Almost 40 ectothermic species have now been tested, and about two-thirds develop fever. Among lizards, fever appears to be confined to one family, Iguanidae. In only a very few of the ectotherms tested has the pyrexia been demonstrated to have survival value. Whether it has survival value in endotherms is still unknown. There are advantages in using antipyretic drugs during fever, particularly because they reduce several unpleasant symptoms. Since short courses of aspirin, and other antipyretic drugs used by doctors, seldom have adverse effects, it seems advisable to continue to use antipyretics during fever, while we gather further information.

1.11 Case studies

The following examples or case studies have been selected to give reasonable coverage of the animal kingdom and at the same time to illustrate some important principles. The reader is encouraged to use the heat balance equation (eqn [1.9]) and the principles discussed under the physics of heat exchange, to evaluate these examples in more detail.

1.11.1 Arthropods

Most arthropods, whether they be aquatic or terrestrial, assume a body temperature under natural field conditions that is close to ambient temperature. There are, however, notable exceptions to this generalisation which we shall discuss later in some detail. Also, even small differences between body and ambient temperatures can be significant in the ecology of insects. For example, Willmer and Unwin (1981), using a set of fine thermocouples and a simple reflectometer in the field, have shown that large insects will reach and maintain higher temperature excesses ($T_b > T_a$) than small insects (Fig. 1.20). Also, if these large insects have a low reflectance they will reach their maximum temperature excess more swiftly than if they were light in colour. Willmer and Unwin (1981) therefore maintain that a knowledge of insect size and reflectance, in association with reliable microclimatic data in a specific

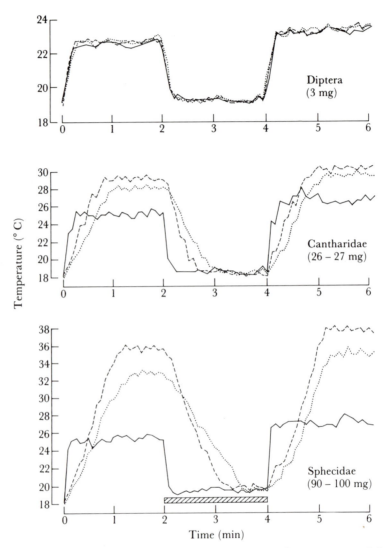

Figure 1.20 Thoracic temperature as a function of time for insects in sunlight (radiant flux 840–900 W m^{-2}) and in shade (hatched bar). The solid lines indicate air temperature, the dashed line the less reflective insect of each pair, and the dotted line the more reflective. The insects were *Xyphosia* sp. and *Hilara* sp. (Diptera), *Rhagonycha fulva* and *Cantharis livida* (Cantharidae), and *Crabro cribrarius* and *Cerceris arenaria* (Sphecidae). Redrawn from Willmer and Unwin (1981).

region, are of considerable predictive value. For example, small insects with a lower temperature excess should be able to reduce body temperature rapidly in the shade, whereas large insects, with higher temperature excesses, are in danger of overheating under hot conditions. In temperate conditions, large insects would be at an advantage but small insects may be obliged to restrict their activities to periods of intense radiation or patches of sunlight. For an excellent and comprehensive review on microclimate and the physiology of insects, the reader is referred to Willmer (1982).

Adaptation to cold

Because of their small size, most arthropods can escape to a favourable microclimate to avoid extremes of climate. These periods of escape are usually associated with dormancy (diapause) during which growth and development are arrested and metabolic rates fall to a minimum. Dormancy can occur during periods of suboptimum low temperatures (hibernation) or superoptimum high temperature (aestivation), or for other reasons, such as lack of water or sufficient food (athermopause). Additional responses to cold include aggregation behaviour, development of increased melanism and hairiness, as well as biochemical changes which can increase metabolic rate by as much as three- to five-fold.

The aggregation behaviour of insects during cold is well illustrated by the honey-bee, and has been re-examined recently by Southwick (1985) and W-Worswick (1987). Philip W-Worswick (1987) exposed honey-bee colonies to decreasing ambient temperatures, measuring their metabolic rate (O_2 consumption) and recording the temperature profiles within the colony simultaneously. He placed a thermocouple grid within the hive and scanned the temperatures of the various points within the grid using a microprocessor. The results (Fig. 1.21) showed that exposure to very low temperatures resulted in an increase in mass specific metabolic rate in the case of Cape bees (*Apis mellifera capensis*) but not in the case of African bees (*A. m. adansonii*). In both subspecies intense clustering occurred at low temperatures but African bees maintain a significantly larger area of the colony at a high temperature (Fig. 1.22). These differences between these subspecies of honey-bees are probably related to differences in their respective habitats. The habitat of the Cape bee is typified by long, cool winters and provides few suitable nesting sites because of the paucity of trees. In contrast, the African honey-bee experiences mild winters and swarms far more frequently in fairly well-wooded savannah habitats.

Bees are not the only insects to raise their temperatures in the cold. It has been known for almost a century, since the discovery of the first thermocouple, that night-flying moths can maintain a body temperature well above ambient temperature in order to produce sufficient power for flight. Recently, Heinrich and Mommsen (1985) found that certain noctuid moths can fly at ambient temperatures as low as 0 °C while maintaining thoracic temperatures of 30–35 °C, an extraordinary achievement for a so-called ectotherm.

But even more extraordinary is Heinrich's finding that geometrid winter

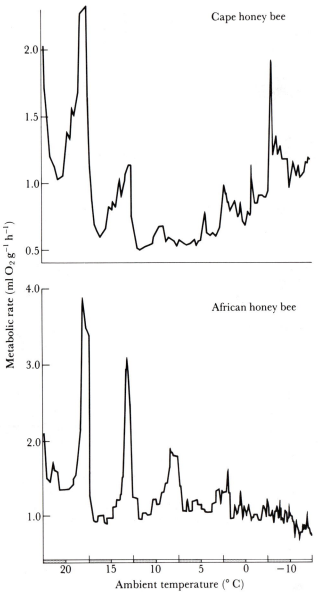

Figure 1.21 The effect of decreasing ambient temperature on the mass-specific metabolic rate (expressed as oxygen consumption) of the Cape honey-bee (upper panel) and the African honey-bee (lower panel). The photoperiod and ambient temperature are shown on the abscissa. Redrawn from W-Worswick (1987).

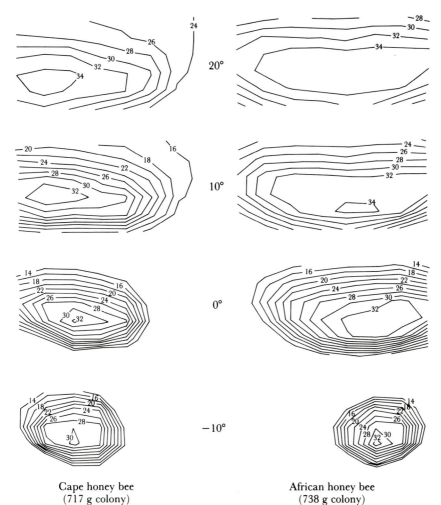

Cape honey bee
(717 g colony)

African honey bee
(738 g colony)

Figure 1.22 Thermal profile through the brood nest of two subspecies of honey-bee at four different ambient temperatures. Redrawn from W-Worswick (1987).

moths can fly at subzero thoracic temperatures. The muscle enzymes of the noctuid and geometrid moths do not differ significantly in their response to temperature (see section 1.1), nor do the metabolic rates of the thoracic muscles at near freezing temperatures. The secret of the success of the geometrids apparently lies in their unusually low wing loading which allows them to fly with minimum power output.

Flying insects, however, more often need high thoracic temperatures during flight to sustain the very high power output which flying requires (Fig. 1.23). For this reason many flying insects cannot fly until ambient

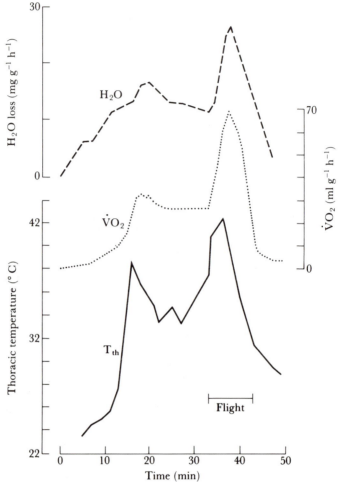

Figure 1.23 Simultaneous measurement of thoracic temperature, evaporative water loss, and oxygen consumption in a carpenter bee during pre-flight thermogenesis and a short flight.

temperatures exceed a threshold; carpenter bees, for example, require a globe temperature of about 23 °C (Louw and Nicolson, 1983). Before attempting to fly, many flying insects will engage in preflight thermogenesis, which consists of shivering (moths) or isometric muscle tension (carpenter bees). Both procedures result in a rapid increase in thoracic temperature (see Fig. 1.23). Shivering in the thoracic muscles of the bumblebee not only allows it to fly at low ambient temperatures (about 3 °C) but allows the queen bumblebees to incubate their brood at nest temperatures of about 30 °C, even when ambient temperature is less than 3 °C. The heat produced in the thorax is transferred to the abdomen by haemolymph moving from the

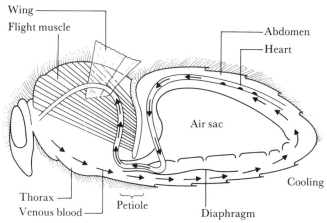

Wing
Flight muscle
Abdomen
Heart
Air sac
Cooling
Thorax
Venous blood
Petiole
Diaphragm

Figure 1.24 The flight muscles of honey-bees and bumblebees are supplied with blood from the heart, which is located in the abdomen. Countercurrent arrangements in the petiole retain heat in the flight muscles. When the bee needs to dissipate excess heat, at high air temperature, the countercurrent mechanism is circumvented by altering the flow pattern thus eliminating countercurrent exchange. Redrawn from Heinrich (1976).

thorax to the abdomen. When the bumblebee is engaged in preflight thermogenesis, however, it restricts the flow of haemolymph to the abdomen. The heat therefore remains in the well-insulated thorax until thoracic temperature becomes excessively high, whereupon haemolymph flow to the abdomen is reinstituted (see Fig. 1.24). In addition to shivering, it is also possible that bumblebees use futile biochemical cycles to produce heat. Bumblebees have relatively large quantities of the enzyme fructose bisphosphatase, which allows the functioning of the futile cycle shown in Fig. 1.25; this enzyme is absent from the thoracic muscle of carpenter bees, which can fly only at ambient temperatures above 23 °C.

When faced with extreme cold (<0 °C) arthropods must either tolerate the freezing of their tissues or avoid freezing by lowering the freezing point of their body fluids and/or employ supercooling. To understand these principles let us first examine a standard cooling curve (Fig. 1.26). The freezing point of an aqueous solution is determined by the concentration and chemical nature of the solutes. Freezing point for the body fluids of most arthropods lies between −0.5 and −0.9 °C. Most solutions will supercool well below the true freezing point in the absence of nuclei which are required for ice crystal formation. When ice crystals do form they can disrupt normal protoplasmic structure, and freezing traps and crystallises salts. Osmotic balances are disturbed and water is drawn out of cells. Also, the diffusion of O_2 and CO_2 is 10^4 times slower in ice than in water.

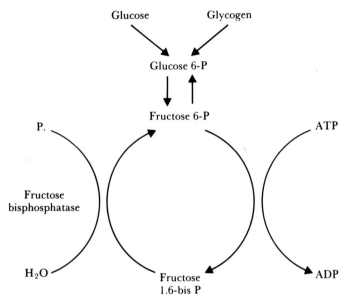

Figure 1.25 Example of a futile biochemical cycle that will waste ATP and produce large amounts of heat.

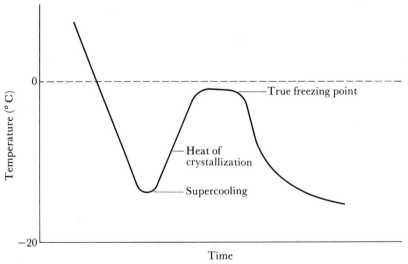

Figure 1.26 A standard cooling curve of an aqueous solution, at the phase interface between liquid and solid.

Insects tolerant to freezing synthesise compounds known as cryoprotectants. The most commonly found are polyhydroxy alcohols such as glycerol, sorbitol and mannitol. They are also very effective in lowering freezing

points. For example, a 5 M solution of a non-dissociating solute will freeze at
$-9.3\,°C$, where a 5 M glycerol solution reduces the freezing point to $-17.5\,°C$.
These compounds act by inhibiting nucleation, the seeding process, by binding to embryonic ice crystals. They may alter the shape of ice crystals by
making them more rounded, thereby reducing injury to plasma membranes.
Hydrogen bonds are also formed between the hydroxyl groups of the cryoprotectants and water. The following data from a study on the parasitic wasp
Bracon, which overwinters in wheat stems, in Manitoba, Canada, illustrates
how effective cryoprotectants can be:

	Supercooling point	*Glycerol*
Summer	$-29\,°C$	0.02 M
Winter	$-47\,°C$	2.80 M

Gut contents are particulate and therefore could promote freezing; consequently, before overwintering, insects evacuate their gut contents. Overwintering insect eggs are also often exposed to extreme cold and some incorpor-
ate cryoprotectants, such as sorbitol.

Adaptation to heat

Arthropods that occur naturally in dry, hot regions are generally capable of
tolerating higher body temperatures than are mesic species; Neil Hadley
(1988) has estimated their upper critical maxima to lie between 45 and 47 °C.
The cause of heat death in arthropods has still not been clearly elucidated
but is thought to involve changes in haemolymph osmotic pressure, decrease
in pH, uric acid accumulation, too little Na^+ and too much K^+; heat-
sensitive enzymes responsible for Na^+ and K^+ transport may also be de-
stroyed. In their natural habitats, however, whenever they can, arthropods
will escape to a favourable microclimate to avoid excessively high body
temperatures. This tactic is well illustrated by a great variety of flightless
tenebrionid beetles that live in the massive sand dunes of the Namib Desert.
The sand surface temperature of these dunes regularly exceeds 60°C,
whereas the beetles maintain body temperatures within the range of 32–
37 °C (Seely *et al.*, 1988), which are within 0.5 °C of the globe temperatures
monitored concomitantly. The beetles therefore behave like inert black
bodies and do not thermoregulate either behaviourally or physiologically. As
soon as the surface temperatures approach their critical maximum temperature, however, the beetles escape beneath the sand to cooler levels. In fact,
Mary Seely and her colleagues maintain that the beetles tolerate high body
temperatures because they have a readily available thermal refuge in the
form of cool sand beneath them (see Fig. 1.27 for profile of sand temperatures).

Many Namib dune beetles are coloured black, which has led to considerable speculation among ecologists and behaviourists. Bill Hamilton (1975)
has advanced the hypothesis that it is to the advantage of many animals to

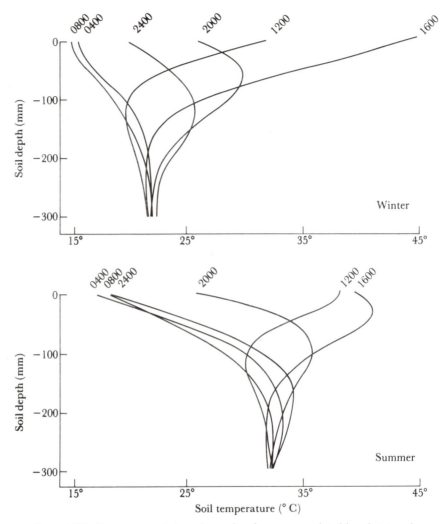

Figure 1.27 Temperatures below the sand surface, measured at 4-hourly intervals
on a dune slipface in the Namib Desert in winter and summer. Note
that there is a benign temperature available below the surface at all
times of day in both seasons, in spite of extremes on the surface. Note
too that the surface temperature extremes are not very different in
summer and winter. Redrawn from Seely *et al.* (1988).

maintain their body temperatures at a maximum level of about 38 °C for as
long as possible ('maxithermy'). By doing so, the animals are also able to
maintain various life processes such as feeding, digestion and metabolism at
peak levels; maxithermy can be advantageous only if there is a reliable food
supply and a readily available thermal refuge. This maxithermy hypothesis

Plate 1.2 A desert ant, *Ocymyrmex barbiger*, standing with a stilt-like posture on a pebble to escape the extreme heat of the surface of the sand. The high surface area to volume ratio of ants allows them to off-load heat rapidly. (Courtesy of A.C. Marsh)

could explain the paradox of black beetles in the hot desert dunes. Before we become too enthusiastic about maxithermy and black beetles, we should remember our earlier discussion about colour (section 1.3.3). Closer examination of the heat exchange in a black and a conveniently similar but white-coloured beetle species in the Namib dunes, tells a different story. Mandy Lombard (1989), in a series of wind-tunnel and field experiments, has demonstrated that the colour of these beetles is irrelevant to all the avenues of heat exchange except direct visible radiation. Because of the overwhelming importance of long-wave re-radiation from the surface of the hot dunes, as well as of convection in the heat exchange of the beetles, she concluded that colour has a negligible influence on their thermal biology. In this case Hamilton's interesting hypothesis did not survive the test of experimentation, but it may well be applicable under different conditions.

In extremely hot regions, there are many advantages to living most of one's life underground and, as we shall see later, to becoming social. Many animal species do so, and they avoid the dangers of overheating entirely by emerging only when favourable microclimatic conditions prevail. It is therefore most surprising to learn from Alan Marsh (1985) that the 4 mg ant species *Ocymyrmex barbiger*, which stands about 4 mm above the surface of the desert sand, forages in full sunlight when the sand surface temperatures are as high as 67 °C. Our knowledge of the physics of heat exchange tells us that an ant as small as this, because of its shape and high surface area to volume

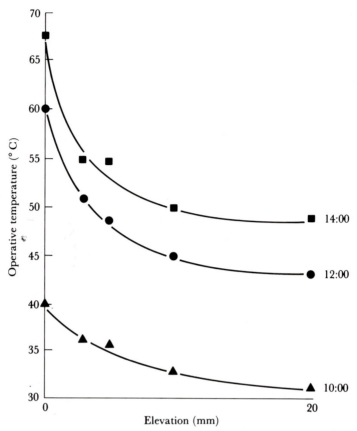

Figure 1.28 Operative temperature affecting ants in the boundary layer above the Namib Desert surface at three times during the day on 31 March 1983. Redrawn from Marsh (1985).

ratio, will off-load heat very rapidly by convection. This advantage, frequently overlooked by biologists, is exactly the tactic which is exploited by the ant. The foragers periodically escape to cooler thermal refuges by seeking out small patches of shade or climbing onto a pebble or dry grass stalk. By doing so they can employ the steep thermal gradient above the surface of the sand; a vertical movement of only a few millimetres can reduce air temperature significantly (Fig. 1.28).

Marsh (1985) found, as one might predict, that the length of pauses within the thermal refuges as well as their frequency increased with increasing sand surface temperatures (Fig. 1.29). Nevertheless, by measuring operative temperature (see section 1.4) he estimated that, in spite of their adaptive behaviour, the foragers would experience body temperatures greater than 52 °C for short periods. In the laboratory he later demonstrated that the critical

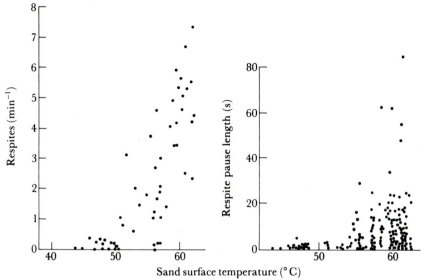

Figure 1.29 Relationship between sand surface temperature, respite frequency and respite pause length for the ant *O. barbiger* foraging in the Namib Desert. Each point represents one event for one individual. Redrawn from Marsh (1985).

thermal maximum for this species was 51.5 °C and that *O. barbiger* could tolerate temperature in excess of the critical thermal maximum for short periods. The question that ecophysiologists should now pose is 'Why indulge in such risky behaviour by walking a thermoregulatory tightrope?' It is a fair question as some of these ants do die while foraging. Marsh's explanation is that by risking foraging at these extreme temperatures *O. barbiger* is the first arthropod scavenger to arrive at the scene of thermal deaths in other arthropods that are 'thermally less astute'.

Similar, but perhaps an even more striking, thermal behaviour has been described by Rüdiger and Sibylle Wehner and Alan Marsh (1989) in the Saharan ant, *Catalglyphis bombycina*. Like *O. barbiger*, *C. bombycina* is a scavenger and feeds primarily on arthropods which have succumbed to heat death on the hot surface of the Saharan plains. Arriving at the scene of a 'heat death' as soon as possible requires foraging at very high surface temperatures which, incidentally, also provide the added benefit of avoiding the ants' major predator, a lizard. The ants tend to forage together, in large numbers, within a narrow surface temperature range, which is just too high for the lizard and just below their own critical thermal maximum. Their brief sortie is illustrated in Fig. 1.30; foraging occurs as a mass eruption or exodus of ants from the nest confined to a period of about 15 min, a remarkable exploitation of a narrow thermal window or thermal niche.

The ants maximise the probability of encountering their prey and

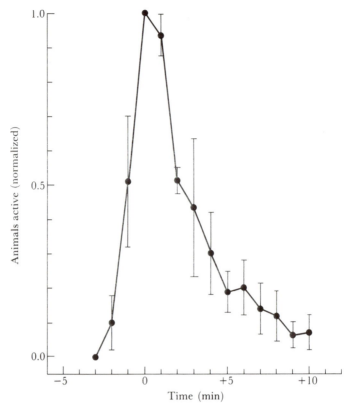

Figure 1.30   Mass exodus and return of foraging *Cataglyphis bombycina* ants over a
13-min period. Redrawn from Wehner (1989).

minimise the probability of encountering their predator, the desert lizard.
This high-risk behaviour and its benefits are only possible because of the
differences in the respective body sizes and surface areas of the ants and the
lizards. The small mass and high relative surface area of the ants allow them
to exchange heat very rapidly, whereas the lizard, with its higher thermal
inertia, is unable to do so. The high heat tolerance of the ants also exposes
the fallacy of claiming that organisms cannot survive at tissue temperatures
above 40–45 °C.

Spiders of *Scothyra* spp. make a capture web on the dune surface (Fig. 1.31)
and will attack insect prey trapped on the web even when the surface temperature is 65 °C. Joh Henschel and Yael Lubin (1990) have discovered that the
spiders behave abnormally when their own temperatures reach 50 °C. They
are able to exploit the much hotter surface because they build tubular silk-
lined burrows in the sand below the web; at the bottom of the burrow,
150 mm below the surface, the sand can be 20 °C cooler than at the surface
(see Fig. 1.27). The spiders shuttle between the bottom of the burrows, where

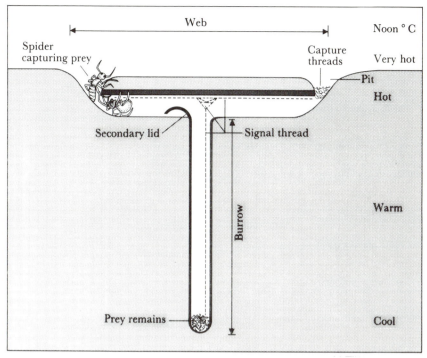

Figure 1.31 Capture web and burrow of the spider *Seothyra*, on a Namib dune. The spider remains below its lethal temperature by retreating to the bottom of the burrow, and making rapid forays when its prey is trapped in the web, which is on the hot sand surface. From Henschel and Lubin (1990), by copyright permission of the Desert Ecological Research Unit of Namibia.

they off-load heat, to the capture webs for brief attacks on the prey. When the surface temperature exceeds 55 °C, prey-handling bouts last no more than a few seconds.

1.11.2 Fish

Most fish species assume the temperature of the surrounding water, not only because of the high heat transfer between body surface and water, but also because of the large surface area of the gills. The gills are highly vascularised, to facilitate O_2 transport from the water to the tissues, and, consequently, any metabolic heat produced in the tissues is carried by the bloodstream to the gills where it is rapidly dissipated by convection into the water. Under certain conditions, however, it is advantageous for fish to maintain body temperatures above that of the ambient water; for example predators hunting swiftly moving prey. No fish can maintain all of its body core appreciably

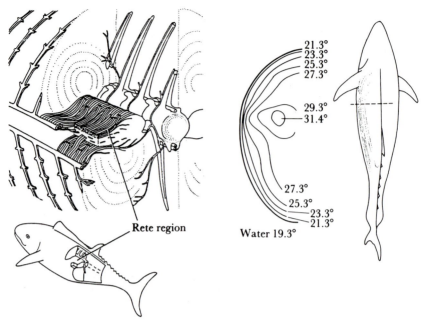

Rete region   Water 19.3°

21.3°
23.3°
25.3°
27.3°

29.3°
31.4°

27.3°
25.3°
23.3°
21.3°

Figure 1.32   Exchange of heat between arterial and venous blood, in a rete, enables the blue-fin tuna to retain heat produced in active deep muscles. Redrawn from Carey and Teal (1966).

warmer than the water, but some can keep certain organs, like muscles, warm. In the case of tuna, muscle heat conservation derives from the development of a countercurrent heat exchanger, in the form of a rete mirabile, between the muscles and the gills (see Fig. 1.32 for explanation). Tuna muscles probably are not kept warm permanently, but are warmed transiently for specific locomotor activity.

Muscles are not the only organs kept warm by fish. The so-called 'warm fish', such as tunas and mackerel sharks, can also maintain brain and eye temperature some 4–15 °C above water temperatures through the conservation of metabolic heat by means of countercurrent heat exchangers in the circulation to the brain and the eyes (Linthicum and Carey, 1972). Even more impressive is the exciting phenomenon discovered by Barbara Block (1987) that in large predators such as marlins and sailfish the brain and eyes are warmed by a specialised thermogenic tissue. This tissue is modified eye muscle tissue, which, like BAT, has large mitochondrial volumes and is richly supplied with capillaries. Unlike BAT, however, it does not produce heat by uncoupling oxidative phosphorylation and reducing ATP synthesis, but appears rather to accelerate the turnover of ATP. Prominent ATPases are situated close to the mitochondria to hydrolyse ATP and, according to

Block, all that is necessary is to cycle calcium across the heater cell membrane system at the expense of ATP. Why has this interesting mechanism evolved? Block concludes that warming of the retina and the visual apparatus may improve the visual acuity of these 'blue water hunters', that have to range over large distances in the open ocean where food is scarce and temperatures can change rapidly. By sustaining their visual acuity, predation success may be enhanced significantly.

The majority of fish species are unable to retain metabolic heat and, consequently, are obliged to survive at ambient water temperature; to do so they employ various behavioural and physiological adaptations. One of the most important of these is the ability to synthesise iso-enzymes that are suited to the temperature regimen to which the fish must adapt. Iso-enzymes (or isozymes), as their name implies, function as catalysts for a particular biochemical reaction but, because of changes in their molecular configuration or geometry, operate optimally in different physical and chemical environments, such as different pH and temperature ranges.

Adaptation to extreme cold in fish has received considerable attention and it appears that polar fish survive subzero temperatures by employing either supercooling or so-called antifreeze compounds in their tissues. Supercooling exposes the fish to the danger of instant death by freezing if they should come into contact with an ice floe. Obviously, this happens very seldom. In polar fish with a low freezing point the normal concentration of Na^+, Cl^-, urea and free amino acids in their body fluids accounts for half the freezing point depression, and the remainder results from an antifreeze compound. The major antifreeze in most of the fish with this adaptation is a glycoprotein. It is a large molecule ($26\,000$–$32\,000$ kDa) and therefore does not contribute very significantly to the osmotic pressure of fish serum, but is thought to envelop embryonic ice crystals which would otherwise act as 'freezing nuclei'. It consists of long chains of the amino acids alanine and threonine attached to the sugars galactose and N-acetylgalactosamine. Arctic fish, however, die from heat stress at comparatively low water temperatures ($<10\,°C$).

1.11.3 Amphibians

As we pointed out previously (section 1.8.2), amphibians are not usually effective thermoregulators. They do not sweat, pant or shiver. Nevertheless, they can exercise a modicum of control over body temperature by behavioural means such as sun-basking and moving in and out of water, and some frogs have access to physiological mechanisms too. Under conditions of heat stress some *Hyperolius* species increase evaporative water loss (Withers *et al.*, 1982). Phil Withers and his colleagues also studied the change in reflectance (colour) these frogs undergo when exposed to high levels of incident shortwave radiation. Reflectance values changed from 0.35 in the normally coloured frog to as high as 0.60 in the blanched frogs. Similar results were obtained by Schmuck *et al.* (1988).

Survival of amphibians in conditions of extreme heat or extreme cold is usually by escape behaviour, known as retraherence. In this way *Rana sylvatica* survives in cold boreal regions of North America where ambient temperatures can reach −50 °C. Similarly, but at the other extreme, *Scaphiopus* the spadefoot toad of North American deserts, and *Pyxicephalus*, the bullfrog of the African savannahs, escape to cool moist depths during unfavourable hot, dry conditions on the surface. Finally, amphibians also employ isozyme adaptations to reduce the impact of ambient temperature upon the metabolism.

1.11.4 Reptiles

Because of the wide distribution of reptiles and their excellent ability to thermoregulate behaviourally, very many thermal studies have been carried out on lizards, snakes and tortoises. I have chosen a few examples to illustrate the flexibility of reptiles in coping with very different thermal environments, and to emphasise certain important principles.

The precision with which lizards can thermoregulate about their selected temperature in the field is well known. Naturally, sources of external heat and cooling must be available to the animals during the period of active regulation. The giant monitor lizards (*Varanus* sp.) are thought by some to be the most efficient thermoregulators among the reptiles; when we placed them in a photothermal gradient, we found that they were able to regulate very precisely at 35 ± 1 °C as long as a heat source was available to them (Louw *et al.*, 1976). These lizards also exhibited the highest oxygen pulse ever recorded for reptiles (i.e. the amount of O_2 consumed per heartbeat) and some biologists feel they represent a transitional phase between ectothermy and endothermy. This view is not really justified when one examines their aerobic scope, and their relatively poor exercise capacity. Moreover, much smaller reptiles such as chameleons are also capable of precise temperature regulation by changing posture, orientation and colour (see section 1.8.1 and Burrage, 1973).

Like amphibians and fish, and indeed many other taxa, including invertebrates, reptiles employ isozymes for temperature compensation. Licht (1964) has established interesting differences in the temperature optima of the enzyme ATPase extracted from the skeletal muscle of different lizard species. The temperature at which peak activity of the isozymes occurs ranks in the same order as the selected body temperatures of the various species from which they were extracted (Fig. 1.33).

We have already seen that the study of reptiles, because they are ectotherms, has allowed important insights into the mechanism of fever (section 1.10). They also provide interesting experimental material for biologists

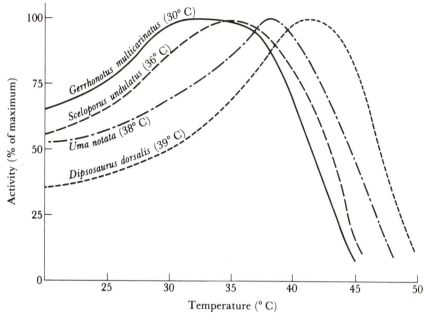

Figure 1.33 Influence of temperature on the activity of ATPase from the skeletal muscles of lizards with different thermal preferenda. Selected body temperatures of the species in laboratory conditions are shown after the species name. Modified from Licht (1964).

interested in the physics of heat exchange. An appreciable literature has accumulated on this subject, and we shall draw attention to just one important question, 'How does body size affect thermal conductance in ectotherms?' Conventional wisdom would have us believe that the ratio of heating and cooling rates in reptiles is solely a function of body size. Scott Turner (1988), however, points out that theoretically there must be an optimum size for the control of heating and cooling rates of reptiles, given the physiological constraints under which these animals operate. This prediction has been confirmed by experiments on alligators (see Fig. 1.34 and Turner and Tracy, 1985).

Most studies of the thermal ecophysiology of reptiles concentrate on their conspicuous surface activity. Yet temperate- and warm-zone reptiles spend most of their time sequestered in refuges, in burrows and under rocks, Raymond Huey and his colleagues (1989) have shown, using garter snakes as an example, that appropriate refuge selection can result in a more favourable body temperature regime than surface activity can achieve. Using radiotelemetry in live gravid female snakes, model snakes, and thermocouple measurement of microclimate temperatures, they showed that, from a wide variety of rocks in their habitat at Eagle Lake, California, the snakes

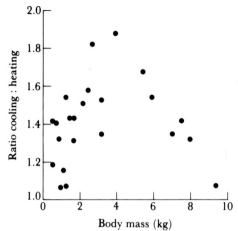

Figure 1.34   Ratio of the cooling rate to the heating rate for alligators weighing between 600 g and nearly 10 kg. The ratio reaches a maximum at a body mass of about 5 kg. From Turner and Tracy (1985).

preferentially selected intermediate-sized rocks (200–400 mm thick) under which to seek refuge. Thinner rocks exposed the snakes to potentially lethal heat stress in the day, and low temperatures at night; thicker rocks did not allow the snakes to warm to their selected body temperatures (Fig. 1.35). By picking a rock of appropriate size, a snake could maintain body temperature within its preferred range for 20 hours of the day, as against 10 hours if it shuttled between sun and shade on the surface.

For a thorough treatment of the thermal biology of reptiles the reader is referred to two recent texts on the ecology and physiological ecology of lizards (Bradshaw, 1986; Pianka, 1986).

1.11.5 Birds

In birds we meet the first true endotherms on the phylogenetic scale. Incidentally, but perhaps not coincidentally, we also meet, for the first time, animals that possess the loop of Henlé in their kidneys and are capable of producing a hyperosmotic urine. Their average heat production is about 8–10 times greater than reptiles and their muscle tissues have a greater concentration of mitochondria. These mitochondria also have a larger relative membrane surface. Birds do not sweat, but do exhibit panting and gular fluttering, a form of evaporative cooling unique to birds, and they are able to shiver. They also employ behavioural thermoregulation and have good control over vasomotor tone.

Response to cold

The body temperatures of most bird species are slightly higher than those of

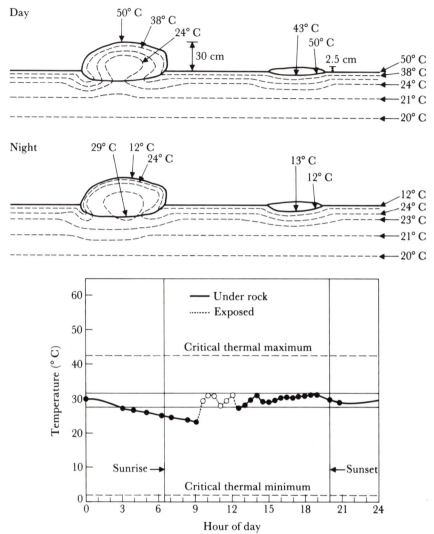

Figure 1.35 The habitat of the garter snake at Eagle Lake, California, contains rocks of various sizes. The upper panel shows the temperature gradients within and below thick (30 cm) and thin (2.5 cm) rocks in the middle of a clear day and of a clear night. Note that, during the day, the temperatures immediately below the rocks are lower than at the rock edges, and that the temperatures under the thin rock are much higher than under the thick rock. During the night, the centre temperatures are now higher than the edge temperatures and the temperatures under the thick rock are much higher than those under the thin rock, or in the soil distant from rocks. By picking a rock of an appropriate size, a snake can maintain its body temperature within its preferred range for most of the day (lower panel); it turns out that intermediate-sized rocks are best. Redrawn from Huey *et al.* (1989).

mammals, within the range of 40–42 °C, and a common response of birds to cold is to migrate to a more favourable environment. Those that remain as residents of cold climates exhibit a range of adaptations. For example, some seek out protected niches in the environment such as caves or hollow trees and the Arctic-adapted ptarmigan even makes use of snow-drifts. Cold-adapted birds increase fat deposition, which is important for insulation and as an energy store. Some species develop a heavier plumage with down-like feathers that trap air very efficiently: the plumage of the ptarmigan is about eight times heavier than that of a similarly sized tropical bird. Additional adaptations include increased metabolic rate (heat production) or the opposite, namely entry into torpor. During torpor, although some birds such as the road runner, will reduce their metabolic rate to approximately one-thirtieth of normal and their heart rate to one-twentieth of normal, they nevertheless maintain constant body temperatures by regulating around a new, much lower set point. Although birds employ torpor, they do not hibernate and to date there is only one example of aestivation, namely the poorwill (Bartholomew, 1982).

Probably the most dramatic example of cold adaptation in birds is provided by the large emperor penguin (*Aptenodytes forsteri*) which not only survives Antarctic winters but actually manages to breed in midwinter. Both sexes of this species leave their normal marine habitat in early winter and walk as far as 100 km inland to the breeding site. The female lays a single egg that is then incubated by the male bird, between the feet and an abdominal fold which partially envelops the egg. After laying, the female returns to the open sea to restore her fat reserves by intense foraging. Meanwhile the incubating male must endure intense cold (−30 to −40 °C) and the chilling effect of the strong Antarctic winds. Not surprisingly the males lose as much as 40% of their body mass during this period of enforced starvation. After approximately two months the female returns to the rookery to feed the newly hatched chick by regurgitating her stomach contents. The male then is free to return to the sea to feed and replenish his fat reserves. How does the male penguin survive these extreme conditions? In the first place, his excellent insulation ensures a low critical temperature well below zero (−10 °C). He probably also employs countercurrent heat exchange in his legs to minimise heat loss to his feet, but at the same time releases sufficient heat for the incubating egg. However, these adaptations are not sufficient to guarantee survival and the crucial adaptation appears to be behavioural, namely huddling (Fig. 1.36). If individual penguins are prevented from huddling with other members of the flock, they lose 0.2 kg of body mass per day, double the rate when huddling. By huddling the birds reduce the surface area exposed to the steep temperature gradient and chilling winds.

In cold environments many bird species are at risk of losing metabolic heat rapidly through their uninsulated legs and feet. Countercurrent heat exchangers in their legs perform the important function of conserving body

Figure 1.36 Chicks of the emperor penguin on their ice-covered Antarctic breeding grounds. The chicks huddle together and thus reduce the surface area exposed to cold air. This behaviour leads to a substantial reduction in the metabolic cost of keeping warm.

heat that would otherwise be rapidly dissipated. Possessing the countercurrent mechanism is particularly important for wading birds, birds standing on ice and during flight under cold conditions. Pete Scholander (1955) demonstrated that a gull will lose only 1.5% of its body heat while standing in ice water for 2 hours. A cross-section of such a bird's leg reveals that the heat exchanger consists of a large artery carrying arterial blood to the extremity, surrounded by many thin-walled veins transporting cold venous blood back to the core of the body. Heat moves from the artery down the temperature gradient to warm the cool venous blood returning to the body, thereby recycling heat back to the core tissues and preventing it from being lost to the environment. The fact that the system has countercurrent flow improves its efficiency by maintaining a temperature gradient throughout its length (see Fig. 1.37).

The heat exchanger in bird legs produces a condition called 'regional heterothermy' along the length of the leg (see Fig. 1.38). An interesting

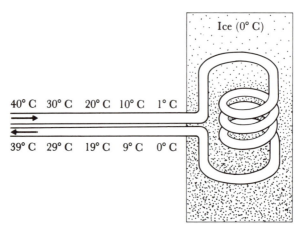

Figure 1.37 Model of a countercurrent heat exchanger such as might exist in the legs of an animal standing on ice. In this case, heat is conducted from the incoming water to the outflowing water, so that in the steady-state condition the outflowing water is pre-warmed to within 1 °C of the incoming water. From Schmidt-Nielsen (1983).

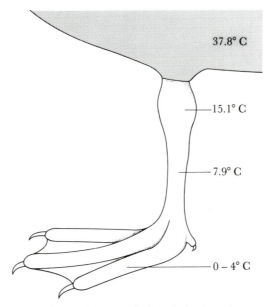

Figure 1.38 Regional heterothermy in the leg of a bird standing on ice.

consequence of this condition is that single cells, for example the neurones of the tibial nerve in the herring gull, which run the entire length of the leg, are exposed to temperatures ranging from 38 °C to 0 °C. Neurones, as we know, are particularly sensitive to temperature and it is remarkable that a single neurone has evolved a conduction mechanism capable of operation at the different temperatures; the threshold temperature for sustaining an action potential of the tibial neurones at the proximal end of the leg is 8–13 °C whereas distally it declines to 2–5 °C (Bartholomew, 1982).

When birds shiver in response to cold, the increase in metabolic heat production can constitute a significant proportion of their total energetic cost of free existence. This observation is not always appreciated by ecologists but it has been amply borne out by the elegant measurements of Paladino and King (1984), who trained white-crowned sparrows to walk on a treadmill at different speeds and at various environmental temperatures, and measured their oxygen consumption. They concluded that it costs a white-crowned sparrow no more energy to move about and forage, when the heat production associated with the exercise keeps the bird warm, than it would cost for it to remain still and shiver to produce the heat necessary to maintain body temperature. These findings have an important bearing on the hypothesis of optimal foraging in birds.

Responses to heat

Because of their capacity for locomotion birds can escape behaviourally from intense heat. This escape usually consists of seeking out shade in appropriate places such as tall trees, restricting activity, and/or exploiting cool air movements to maximise convective cooling. Although birds perhaps are not as proficient as small mammals in finding niches with a favourable microclimate, there are some notable exceptions such as elf owls, gila woodpeckers and rock wrens, which make use of sheltered cavities in the Sonoran Desert. In hot environments, soaring birds can fly to higher altitudes where ambient temperatures are much lower and secondary long-wave radiation is greatly reduced. Another advantage of efficient locomotion is the ability to fly long distances to reach water, which in turn allows extensive use of evaporative cooling under conditions of heat stress.

In general, however, birds tend to rely on their plumage as a barrier to radiation, and on the use of thermal panting and/or gular fluttering, when faced with excessive heat loads. We have already seen that the tips of the desert raven's black feathers can reach temperatures as high as 80 °C, thereby reversing the thermal gradient towards the environment. Gular fluttering, rapid pumping of air to and fro across the wet surfaces of the mouth and throat to amplify evaporative cooling, can reach frequencies as high as 1000 oscillations min^{-1}. Birds also practise selective brain cooling, i.e. the cooling of the brain below the temperature of arterial blood leaving the heart, by means of heat exchange in the ophthalmic rete. This rete consists of a network of arteries and veins in the head. The arteries are the main blood supply to the brain. Evaporative and convective cooling takes place in various regions of the head such as the eyes, upper respiratory tract and, in certain species of bird, bare patches on the head which cool the venous blood entering the rete. Heat exchange in the rete, in turn, cools the arterial blood destined for the brain. Differences of 1.5 °C between brain and body temperature have been measured (Kilgore *et al.*, 1979; Mitchell *et al.*, 1987). This cooling via the ophthalmic rete is analogous to that occurring via the carotid rete in mammals, which will be discussed in section 1.11.7.

Their physiological and behavioural adaptations allow birds to invade almost all habitats and this is amply illustrated by the success of ravens in Middle Eastern deserts and the ice fields of the Arctic. However, very small birds, like humming-birds, are constrained by their surface-to-mass ratios to temperate and tropical environments. Very large birds, and particularly the large flightless birds (ostrich, emu, rhea) do well in open grasslands and deserts. For example, the ostrich, *Struthio camelus*, is too large to escape to a favourable micro-environment. It does, however, have the advantage of high thermal inertia and is able to survive in hot, arid regions throughout Africa and the Middle East. The plumage of the ostrich is manipulated very effectively to facilitate thermoregulation. The feathers are long and sparsely distributed over the dorsal surface of the bird. When ostriches are exposed to high ambient temperature and/or intense radiation, they erect the feathers on the back, thereby increasing the thickness of the thermal shield between the incident radiation and the skin. The sparse distribution of these feathers allows lateral air movement to pass between the feathers and convectively cool a large surface area of skin. In addition, the birds orientate towards the sun and move their wings downwards and away from the thorax. This manoeuvre both shades and exposes the thorax which, because of its naked surface, now acts as a large thermal window for radiant and convective heat loss.

The use of feathers to reduce radiant heat load and enhance convection represents an important water-saving adaptation under desert conditions. On still, hot days, however, the birds are forced to resort to rapid, shallow thermal panting (respiration rate rises suddenly from 4 to about 40 respirations min^{-1}). At night, when the desert air can cool rapidly, the birds are faced with the opposite problem of conserving heat. This problem they solve by folding their wings close to the thorax, interlocking the dorsal feathers and tucking their naked legs beneath them as they huddle close to the ground. In this position the plumage traps an insulating layer of air around the skin, and the temperature of the air space between the skin and the feathers rises to a uniformly high level. The heart rate declines under these conditions, indicating a low metabolic rate and conservation of energy (see Fig. 1.39 and Louw *et al.*, 1969).

When considering the thermal biology of animals, the thermal relations of non-adult phases are frequently overlooked. One is inclined to forget that fertile eggs are living organisms and, in view of the crucial role they play in the survival of so many species, they certainly deserve more attention. The

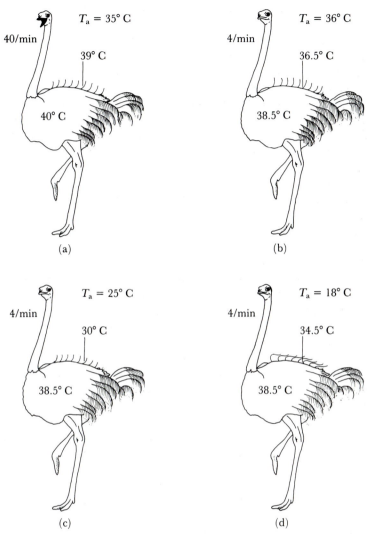

$T_a = 35°$ C

40/min

39° C

40° C

(a)

$T_a = 36°$ C

4/min

36.5° C

38.5° C

(b)

$T_a = 25°$ C

4/min

30° C

38.5° C

(c)

$T_a = 18°$ C

4/min

34.5° C

38.5° C

(d)

Figure 1.39 Thermoregulation in the ostrich involves thermal panting **40** respirations min^{-1} and feather erection when there is no wind and ambient temperature is high (a). In windy conditions feather erection alone is frequently sufficient to allow body cooling (b, c), while at low ambient temperature at night, the feathers are flattened to provide an insulating layer of air (d). Redrawn from Louw *et al.* (1969).

Plate 1.3 The huge communal nests of the sociable weaver, *Philetairus socius* dampen fluctuations in ambient temperature, thereby protecting these birds and their chicks from the extreme heat and cold of the more arid areas of the African savannah.

almost uniform spherical shape of birds' eggs has, however, endeared them to biophysicists interested in scaling the physics of heat exchange in eggs of different sizes (Turner, 1988). Ecologists have also studied the cost of keeping eggs warm under cold conditions, in the context of the insulative properties of nests (Mugaas and King, 1981). Turner's analysis of heat exchange in eggs touches on some very fundamental principles of scaling in animals generally and is highly recommended for further serious reading (Turner, 1988).

Desert birds, unlike their temperate counterparts, face the problem of cooling their eggs. The ostrich lays its eggs in a shallow scrape on the surface of the desert and then relies on its high thermal inertia and its thermoregulatory mechanisms, described previously, to keep the nest from overheating. At high noon in the Namib Desert one can observe ostriches sitting on their eggs in full sun with feathers erected and panting rapidly at 40 respirations min^{-1}. If a cool breeze should begin to blow the birds will frequently orientate in its direction and raise their bodies above the eggs by sitting on their hocks. In this way they continue to shade the eggs but simultaneously expose them to convective cooling. These manoeuvres are obviously successful and even allow the ostrich to breed during midsummer.

A complete contrast to the scrape nest of the ostrich is a most unusual form of nest construction used by desert birds, namely the enormous communal nest constructed by sociable weavers in the branches of *Acacia* trees. This species is common in the Kalahari Desert and invades the semi-arid steppe on the eastern edge of the Namib Desert. Its range is limited by the requirement for *Acacia* trees (Plate 1.3) and suitable grasses for the construction of the nest, though its range has been extended artificially by the erection of telephone poles in the desert, which the birds find quite acceptable. George Bartholomew and his colleagues (1976) examined the thermal biology of sociable weavers and found that the insulation provided by the elaborate nest dampened the thermal consequences of the widely fluctuating ambient temperatures which are so characteristic of the desert environment. The birds are protected from the extreme cold at night and excessive heat during the day. Unfortunately for the weavers, however, this favourable thermal environment is also enjoyed by a large species of cobra which preys heavily on the birds.

1.11.6 Small mammals

Small body size imposes distinct disadvantages on mammals exposed to both heat and cold stress. When faced with cold, small mammals are at a decided disadvantage because there is clearly a limit to the thickness of fur or fat insulation that they will be able to carry (see Fig. 1.40). A very thick fur would make locomotion impossible. In addition, they lose heat rapidly because of their high relative surface area. For these reasons small mammals

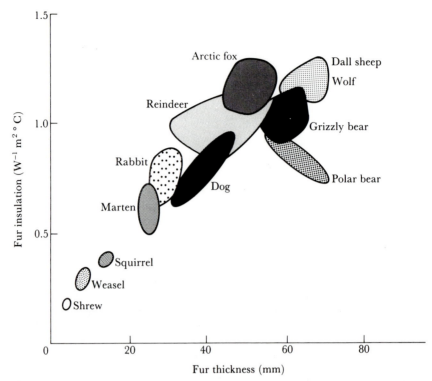

Figure 1.40   Insulation offered by animal fur as a function of fur thickness. The small mammals, which of necessity carry short fur, have poor insulation compared with the larger mammals; fur thickness tends to increase as body mass increases. Compare these cold-environment animals with the African ungulates (Fig. 1.43), where pelage thickness decreases as body mass increases. Redrawn from Scholander *et al.* (1950b).

rely heavily on behavioural escape to protected nests or burrows, where they will frequently roll themselves into as near spherical a shape as possible, or huddle with nest mates and engage in either torpor or hibernation. Many small mammal species rely on BAT for producing large amounts of heat during emergence from torpor or hibernation. Figure 1.40 illustrates the principle that fur thickness increases with increasing body mass in Arctic animals and confirms the relatively thin insulation of small mammals. The opposite situation holds true for ungulates in the warm African savannah where fur thickness declines exponentially with increasing body mass (see Fig. 1.43). The implications of the latter relationship will be discussed presently.

In spite of the disadvantages facing small mammals when exposed to extreme cold, they are surprisingly successful if given sufficient protection and sufficient food. Barnett and Little (1965), for example, reared 19 success-

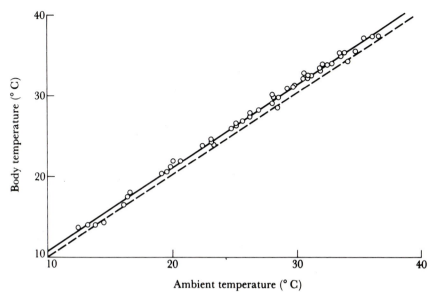

Figure 1.41 Body temperatures of 54 naked mole rats, *Heterocephalus glaber*, measured by Rochelle Buffenstein and Shlomo Yahev, at a variety of ambient temperatures. The solid line is the regression line, and the dashed line the line of identity of body temperature and ambient temperature. These mammals are completely poikilothermic.

ive generations of mice at −3 °C. The metabolic rate of the cold-reared animals was four times greater than the controls kept at 21 °C, they had fewer young than the controls in their first litters and there was a higher mortality among their nestlings. Cold-reared animals consumed about 70% more food than the controls. With an increasing number of generations exposed to cold, the mice produced a greater number of second litters, with a diminishing mortality rate, and maintained their weight better during lactation. The offspring of mice that had been exposed for many generations to cold utilised their food more efficiently than did naïve animals.

If a mammal is not only small but also without fur, it has little chance of being able to maintain body temperature in the cold. Such is the case with naked mole rats (*Heterocephalus glaber*); they are entirely without any insulation in the form of fur. It is no surprise then to learn that they live out their entire lives in warm underground burrows in the hot, arid regions of Kenya. Their body temperatures are identical to burrow temperature; thus they are poikilothermic mammals (Fig. 1.41).

Because of their unfavourable surface area to volume ratio, small mammals, poorly adapted to cold, are also not well adapted to extreme heat. They cannot afford to evaporate water to maintain body temperatures as they would lose too great a proportion of their body water in the process. Small mammals rely almost exclusively on escape behaviour to cool burrows,

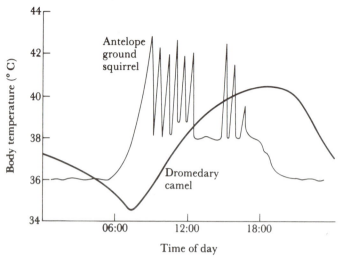

Figure 1.42 Diagrammatic representation of the changes in body temperature, with time of day, of a large and a small mammal subjected to hot desert conditions. Redrawn from Bartholomew (1982).

crevices, etc. when faced with excessive heat. This escape behaviour may be permanent (mole rats), diurnal (kangaroo rats and gerbils), for a few hours at midday (hyrax), or periodically throughout the day to relieve transient episodes of hyperthermia (ground squirrel). It may also be accompanied by a circadian torpor or longer period of aestivation. Figure 1.42 compares the body temperature of the antelope ground squirrel with that of a camel employing adaptive hyperthermia in a similar thermal environment.

1.11.7 Large mammals

Large mammals enjoy several thermal advantages over small mammals. Because of their large size they have a relatively smaller surface area for either absorbing or losing heat. The larger size also provides them with a greater thermal inertia, which reduces the variations in body temperature during storage or loss of heat. They are also able to carry a thick dense pelage, advantageous in cold habitats.

A relatively small surface area is naturally not always an advantage (e.g. when sprinting under hot conditions) but large size is associated with efficient locomotion and this means that large animals can move over long distances to find more favourable habitats or merely to obtain daily drinking water for eventual evaporative cooling. Being large also means that the animals are further from the hot surface of the desert sands and therefore are exposed to much lower temperatures than small animals.

Responses to cold

The most typical adaptation to extreme cold is increased insulation, which is usually accomplished by the seasonal growth of a thick winter pelage, with a specialised fine undercoat. Pelages can double in thickness between summer and winter. The environmental stimulus for the growth of the winter pelage is provided by the photoperiod, which alters the function of the anterior pituitary gland. These thick pelages are very effective in reducing conductance, so much so that sled dogs have been seen to sleep on top of their kennels (à la Snoopy) during blizzards. Reduced conductance can, however, be an embarrassment on occasion. For example, when large mammals such as moose or reindeer are required to sprint away from predators, particularly in deep snow, they are unable to dissipate the resulting metabolic heat adequately through the pelage, even at air temperatures below −40 °C. How reindeer cope with the problem of needing a thick pelage to survive the bitter Arctic cold, yet having to run to escape, has been studied painstakingly by Arnoldus Schytte Blix, Helge Kreuzer Johnsen and their colleagues, using reindeer trained to run on treadmills at the Institute of Arctic Biology in Norway. Reindeer have probably the most sophisticated of nasal heat exchanges. Not only do they employ a carotid rete to keep the brain cool (see below) but they also have a countercurrent blood supply to the turbinate bones, from which they evaporate water into inspired air. When they need to conserve heat, they implement the countercurrent flow, which causes the outer parts of the respiratory tract to drop their temperature to near freezing, so that all the heat and water added to air in the lower tract is retrieved. When the reindeer run, and especially if they pant, the countercurrent system is bypassed, the outer turbinates warm up, and heat loss is maximised. Even with this elaborate system, reindeer remain susceptible to heat stress when they have their winter pelage, and are unwilling to exercise if their body temperatures rise even slightly.

Predators of Arctic mammals, like wolves, also face the dilemma of needing a thick pelage and having to exercise. Though they do not have as efficient a means of brain cooling, they do pant effectively. Also, they may eat snow, like sled dogs, which would serve the dual function of reducing body temperature and slaking their thirst (Folk, 1974).

Aquatic mammals, such as seals, in cold regions have to overcome the special problem of being exposed to near freezing sea water. They rely on reducing their conductance with a subcutaneous layer of fat or blubber, which is a poor conductor. By increasing blood flow to the surface of the body, these animals can increase heat loss to the surroundings. When vasoconstriction reduces the blood flow to the surface, however, the skin and superficial layers of blubber become cold while the deeper layers of blubber remain warm and heat is retained inside this cold shell. Heat loss to the extremities is greatly reduced by the operation of countercurrent heat exchangers, proximal to the flippers. During strenuous swimming, the heat exchangers are bypassed by arterial vasodilation and the return of cool blood by alternative venous routes. The high rate of convection to the water ensures that heat is effectively removed. However, when seals are forced to exercise strenuously on land, their heat transfer to air is too low and they may overheat and die (so-called roadskins).

Naked marine mammals have to rely entirely on the 'cold shell' effect of the blubber layer to conserve heat, but those with fur have the added advantage of trapping a stagnant water layer close to the skin (wet suit effect) which reduces convective heat loss very effectively. Others possess an under-fur that is actually water repellent and traps stagnant air close to the skin with obvious advantages.

Animal fats (like butter) tend to lose their plasticity at near-freezing temperatures. The fats deposited in the peripheral tissues of Arctic and Antarctic animals therefore have short hydrocarbon chains and have melting points some 30 °C lower than fats deposited in the animals' core areas. They keep the soles of the feet of wolves and Arctic foxes supple and flexible at temperatures of -1 °C.

In environments which subject adult mammals to severe cold stress, their neonates, with high relative surface areas, require heroic measures to survive. The 8 kg calves of musk oxen, born in Greenland and northern Canada in the spring, can tolerate ambient temperatures of -35 °C, maintaining a gradient of almost 75 °C between environment and body core (Blix et al., 1984). They use two mechanisms to maintain body temperature. First, their fur, which in the newborn extends over the legs, affords excellent insulation, provided it is dry. Secondly, the calves possess vast deposits of BAT, with an especially high capacity for heat production. This allows them to increase their metabolic rate by more than 50% without shivering, to a value 13 times higher than the human basal metabolic rate.

Responses to heat

Large mammals experience increasing difficulty in losing heat through their surfaces as they become larger, a thermal disadvantage usually well compensated for by their increased thermal inertia. A thick pelage would compromise surface heat loss even further, so it is not surprising that as body mass increases in the ungulates of the warm African savannah the thickness of their fur declines (Fig. 1.43).

This relationship is the exact opposite of the relationship of these variables in Arctic animals, and Hofmeyer and Louw (1987) interpreted it as reflecting an adaptation for facilitating heat loss. An equally valid interpretation would be that smaller mammals, because of their great relative surface area, require more protection against cold. The two explanations are not mutually exclusive.

In view of the dramatic effect of exercise on heat production, any reduction in activity would be most effective in preventing heat stress. This phenom-

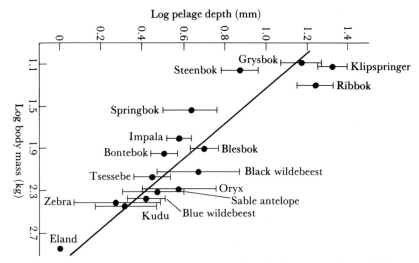

Figure 1.43 Relationship between pelage depth and body mass in selected African ungulates. Compare with cold-environment mammals (Fig. 1.40). Redrawn from Hofmeyr and Louw (1987).

enon is obvious to the most casual observer who ventures out in the desert or savannah in the heat of the day and encounters very few active animals apart from eccentric biologists.

Large mammals employ all the channels of heat transfer they have available in order to keep cool. Conduction of excess heat can be enhanced by seeking out a cool surface, adopting a spread-eagled posture and adpressing the body to a cool surface. In these conditions vasodilation of the superficial blood vessels would increase the rate of heat loss, as would wallowing in mud—a favourite pastime of hippopotami and pigs. Pigs have no functional sweat glands and the common English expression 'sweating like a pig' has no scientific foundation.

If the air temperature is below that of the animal's skin surface, then forced convection will increase rate of heat loss appreciably, depending upon the temperature gradient and the wind speed. Consequently, some animals will actively seek out the slightest breeze on hill tops, such as rhinos in the African savannah, or the desert oryx that moves to the crest of dunes during the heat of the day. Remember that wind speed increases with increasing height above the ground, as the boundary layer effects diminish, with obvious implications for large mammals just as for small arthropods. In section 1.8 we established the importance of colour and orientation of the long axis of the body in reducing profile area and consequently heat gains by radiation in the springbok. This behavioural mechanism is employed by several other large mammals (e.g. wildebeest).

Large mammals can employ conduction and convection to dissipate metabolic and radiant heat provided their thermal environments are within their thermoneutral zones. Once the thermoneutral zone of the animal in question is exceeded, the animal must either store the heat or resort to evaporative cooling. Physiological enhancement of evaporation takes the form of sweating or panting. Naturally, 1 g of water evaporated removes the same amount of heat (about 2.4 kJ) from the animal, whether it is lost by panting or sweating. However, if we examine the comparative efficiency of sweating and panting we note that sweating requires very little energy. If sweat is secreted on to a skin surface which is covered by a fairly thick pelage, the air space between the skin surface and the surface of the pelage rapidly acquires a high water vapour pressure and the evaporation rate is reduced. Also, sweat secreted on to a naked surface cools that surface, which means that on hot days the temperature gradient between ambient air and skin surface is increased, thereby actually increasing movement of heat from the environment into the animal. Panting, on the other hand, requires work to be performed, mostly by the intercostal muscles, producing more heat. The work is minimised, however, in those animals in which panting occurs at the natural resonant frequency of the intercostal muscles, so that the stored elastic energy in these muscles is exploited. Another potential danger of panting, namely the excessive ventilation of the lungs and the consequent development of respiratory alkalosis, is overcome in panting animals by confining the increased ventilation to the respiratory dead space and not the alveoli, where gas exchange takes place (Fig. 1.44). When panting animals exercise, however, they have to increase both dead space and alveolar ventilation. How they control their respiratory muscles to meet the demands of cooling, gas exchange, and pH control is not yet known.

Endurance runners, such as man and the equids, sweat profusely and do not pant, but canids are also outstanding endurance runners and are incapable of sweating. If humans exhibited thermal panting, lectures would be cancelled on hot days and playing the flute in the symphony on hot nights could become an embarrassment.

In many large mammal species, panting has another important advantage in that it cools the mucosal surfaces of the nasal sinuses and consequently the turbinate bones. Dick Taylor (1966), Mary Ann Baker and James Hayward (1967), were the first to point out that cooled blood returning via the carotid rete, situated just beneath the brain, would cool the warm arterial blood destined for the brain. Several workers have since confirmed that the carotid rete is capable of maintaining a temperature difference of 0.5–2 °C between core body temperature and brain temperature (see Fig. 1.45).

The selective cooling effect of the carotid rete on the brain may be important when animals experience mild to moderate hyperthermia even at rest. A recent comprehensive review on selective brain cooling (Mitchell *et al.*, 1987), however, inclines to the opinion that cooling by the carotid rete is especially important during vigorous exercise when work causes a rapid rise

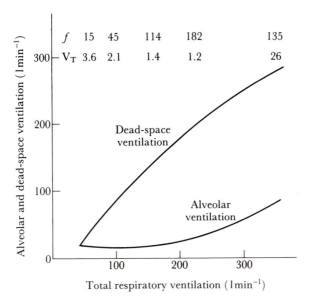

Figure 1.44 With progressively increasing heat stress, total respiratory ventilation increases about seven-fold in the panting ox. The increase derives almost entirely from dead-space ventilation until the total ventilation exceeds about 200 l min^{-1}. Under extreme stress, panting rate (f) decreases and tidal volume (V_T) increases. Redrawn from Hales (1966).

in body temperature. If the animal is able to store heat during exercise without resorting to evaporative cooling, it will save a considerable amount of tissue water, which may be critical under conditions of water stress. Naturally, selective cooling of the vulnerable brain tissue in these circumstances would be a great advantage. To emphasise this point one need only examine the rate of water evaporation required to dissipate the metabolic heat of running in terrestrial quadrupeds of various masses (Fig. 1.46).

The suggestion that the function of the carotid rete is particularly important during the heat loads experienced during and after running is also borne out by the disjunct distribution of the rete among mammals subject to heat stress (Table 1.1). It would seem that if an animal possesses a fur coat and is obliged either to sprint from a predator or after prey, it requires a carotid rete to selectively cool the brain. Alternatively, profuse sweating must be employed.

There is as yet no certainty on how the cooling effect of the carotid rete could be controlled in a graded or stepwise fashion. Mitchell *et al.* (1987) in their review conclude that the most appropriate site for control is between the evaporative surfaces and the rete (Fig. 1.47). In summary then, large mammals, although they are not able to escape into burrows and crevices

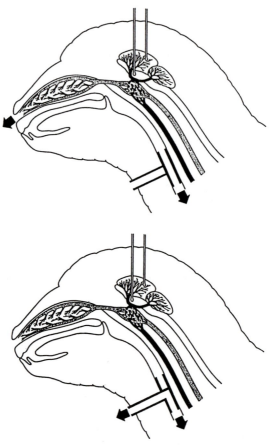

Figure 1.45   The carotid rete, lying just below the brain of certain mammals, allows heat exchange between arterial blood destined for the brain and cool venous blood draining the nasal mucosa. The efficacy of the rete system in maintaining selective brain cooling can be demonstrated by implantation of a tracheostomy cannula, allowing reversible bypass of the upper respiratory tract. Evaporative cooling of the nasal mucosa is abolished when the sheep breathes through the tracheostomy cannula, and thermocouples implanted in its brain show immediate abolition of the selective brain cooling. From Laburn *et al.* (1988).

like arthropods and small mammals, are still able to tolerate extreme heat by various means, of which the following are the most important:

1. By evaporative cooling. Most large animals that have regular access to water can survive extreme heat well. This includes man.
2. By storing heat. This phenomenon is known as adaptive hyperthermia and is best exemplified by the camel (Schmidt-Nielsen *et al.*, 1957) and

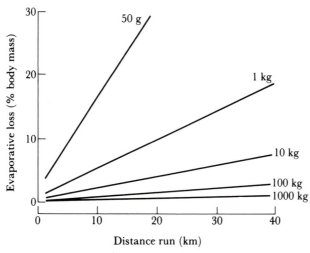

Figure 1.46 Water evaporation, expressed as a percentage of body mass, necessary to dissipate completely the metabolic heat generated in running different distances, for terrestrial quadrupeds of various masses. The potential desiccation of small mammals is so great that they are usually precluded from using evaporative cooling during exercise, and so are largely nocturnal. Very large mammals hardly desiccate at all. Modified from Mitchell *et al.* (1987).

Table 1.1. The disjunct distribution of the carotid rete among mammals.

Group	Carotid rete	Type of locomotion
Camelidae	Yes	Capable of galloping under hot conditions
All true ruminants	Yes	Most capable of galloping under hot conditions (typical prey animals)
Suidae (pigs)	Yes	No functional sweat glands—obliged to sprint from predators
Canidae	Yes	Swift endurance runners—predators
Felidae	Yes	Swift predators (mostly)
Viveridae	No	Not required to run long distances (nocturnal)
Mustelidae	No	Do not run long distances
Rodentia	No	Incapable of running long distances, especially under hot conditions
Equidae	No	Profuse and efficient sweaters/good endurance runner
Humans	No	Profuse and efficient sweater/good endurance runner
Other primates	No	Mostly arboreal and seldom if ever have to sprint from predators for any length of time

(a)

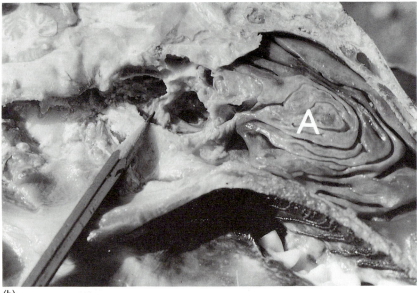

(b)

Plate I.4   (a) Although cheetahs are the fastest sprinters among terrestrial animals, they have poor endurance and cease running when their body temperatures rise to a threshold level. (b) While sprinting their brains are kept cool by a heat exchanger at the base of the brain, the carotid rete (pencil point). Note the elaborate architecture of the turbinate bones in the nasal sinus (A), designed to cool the venous blood returning to the carotid rete.

the oryx (Taylor, 1969). These animals exploit their high thermal inertia, especially when dehydrated, by allowing their body temperatures to fluctuate widely on a diurnal basis (Fig. 1.48). Obviously, selective cooling of the brain by the carotid rete would be of great advantage during adaptive hyperthermia. The net benefit of this adaptation is to

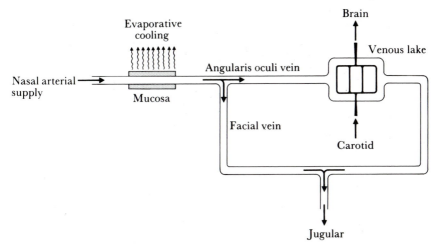

Figure 1.47 Diagram of the cranial vascular architecture which allows blood cooled by evaporation at the nasal mucosa to irrigate, or to bypass, the carotid rete, so altering the degree of selective brain cooling. Based on studies by Johnsen *et al.* (1985) in reindeer. From Mitchell *et al.* (1987).

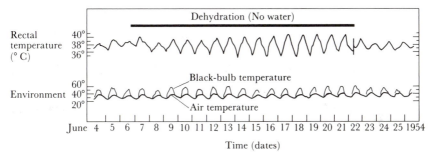

Figure 1.48 Fluctuations in the body temperatures of camels deprived of water or given water *ad libitum*. The greater the dehydration, the greater the diurnal body heat storage. Redrawn from Schmidt-Nielsen *et al.* (1957).

save significant amounts of water which would otherwise be squandered on evaporative cooling to maintain a constant body temperature.

3. Orientation of the long axis of the body to reduce profile areas exposed to direct radiation.
4. Exploiting convective cooling by moving to hill tops.
5. Reducing activity and seeking of shade during the day.
6. Using the pelage as a thermal shield (only certain species, e.g. the dark tips of the mature fleece of a merino sheep can reach temperatures as

(a)

high as 80 °C, thereby reversing the heat gradient to the environment; the air space within the fleece retards flow of heat to the skin surface).

1.12 Concluding remarks

From the foregoing discussion on thermoregulation, it is clear that environmental temperature has a profound effect on life forms and, more particularly, on life processes. The effect is clearly reflected by the decline in species richness from the tropics towards the polar regions, although these differences may merely be due to the greater climatic stability in the tropics, which would allow more time for geographical isolation and speciation. In spite of this profound effect of temperature, gross physiological differences between high temperature and low temperature species are not always apparent and one should guard against simplistic thermophysiological explanations for the distribution of animals. For example, the reduction in the number of bird species that occurs with increasing altitude in montane areas is usually explained on the basis of low temperatures, the resultant high metabolic rates and lack of food. In point of fact, plant cover is reduced at high altitudes which leads to increased predation pressure, and the interaction between predation and diminished food supply is frequently the decisive reason for the sharp decline in species richness (Grubb and Whittaker, 1988).

The apparent lack of gross physiological differences between high and low temperature species can perhaps be best illustrated by taking some comparisons to absurd lengths. We could ask, for example, why there are no

(b)

Plate 1.5 Two widely different ways of keeping cool. The African elephants (a) use their large ears as thermal windows to lose heat and the ground squirrel (b) *Xerus inauris* uses its tail as a parasol to reduce the radiation load from the sun.

crocodiles and hippopotami at the North Pole. In the case of the hippopotamus, there are no major thermoregulatory constraints denying the species access to the Arctic Circle. It is a large endotherm with a relatively small surface area and, if provided with a layer of subcutaneous fat and a fur coat, which are presumably relatively minor adaptations, it would be well suited to the Arctic from a thermal point of view, provided it did not open its large mouth which would allow a great deal of heat to escape. From an evolutionary viewpoint, the reason for the absence of hippos from the Arctic is therefore not necessarily due to insurmountable thermoregulatory constraints, but rather due to nutritional reasons. The North Pole cannot accommodate megaherbivores! In the case of the crocodile, there are some real thermal constraints as it is an ectotherm which relies on solar radiation to thermoregulate. Moreover, its large body size would endow it with a considerable amount of thermal inertia, thereby exacerbating the consequences of the low temperatures and lack of intense radiation. Nevertheless, the crocodile could presumably evolve cryoprotectants and a suitable suite of isozymes, to allow it to be active at the low temperatures and to feed upon Arctic teleosts. Alternatively, it could evolve into an endotherm with suitable insulation. In both instances it would have to compete with large, terrestrial endotherms (polar bears) and aquatic endotherms (seals) that are far better adapted to this niche in terms of morphology, locomotion and nutritional physiology than is the crocodile.

An equally bizarre but, in this case, factual example is provided by the bacteria found in the superheated water (300 °C) surrounding the hydrothermal vents on the sea floor more than 2000 m below the surface of the ocean. At this extremely high temperature one would expect the enzymes and other proteins of these bacteria to be completely denatured, yet these organisms only die when 'cooled down' to the normal boiling point of water (100 °C). Equally remarkable, and more pertinent to our present discussion, is that these bacteria essentially possess the same ultrastructure as normal bacteria when viewed through the electron microscope (Thomas, 1985).

A more realistic comparison of high and low temperature species is provided by comparing an African antelope such as the springbok with the reindeer or caribou. Both are ruminant herbivores, possess a very similar general morphology, yet they inhabit vastly different environments, namely the African savannah and the Arctic Circle. The only major difference between the two species, however, is the superior insulation possessed by the reindeer in the form of a thick pelage. In contrast, the springbok, which is faced with excessive heat loads, particularly when sprinting away from its major predator, the cheetah, has almost no subcutaneous fat and only a very thin pelage. This poor insulation facilitates heat loss under these and other high-temperature conditions, whereas the superior insulation of the reindeer only becomes an embarrassment when sprinting away from its major predator, the wolf, especially when running through deep snow. In the latter case, the excessive heat load is dissipated by panting and the brain is protected by heat exchange in the carotid rete. The brain of the springbok is similarly protected when engaged in sprinting and it is merely the gradient between the environmental temperature and the core temperature of these two species that has imposed either poor or superior insulation upon their morphology.

It is also significant to note that humans can tolerate environments ranging from the tropics to the polar areas merely by altering their insulation. Many similar comparisons could be made. For example, Larry Slobodkin has emphasised that desert organisms utilise the same basic physiological machinery as their temperate relatives and any physiological modifications that may exist have evolved to cope with either the episodic rich food resources or continuously low quality food resources that characterise deserts (Dobson and Crawley, 1987).

The above examples used to illustrate the similarity of animals in terms of their basic thermoregulatory physiology and the caution which should be employed when explaining animal distribution in terms of thermoregulatory constraints should, however, not be taken too far. For example, small ectotherms are ideally suited to desert areas, hence the abundance and variety of arthropods and small reptiles in these regions. By behavioural means, they can escape extremely high temperatures and at the same time, when necessary, soak up solar radiation thereby avoiding the necessity for using nutrients to produce heat endothermically. Also, fascinating differences exist in the isozyme complements of similar animals occupying very different thermal niches. For example, we have seen how desert ants can survive transient temperatures in excess of 50 °C as they exploit their small body size to off-load heat rapidly by behavioural means. These insects also illustrate that shape is almost as important as size when analysing heat flux through organisms, because the ant's filament-like geometry minimises radiant heat gain and maximises heat loss. Conversely, heat loss is minimised by a blocky, rotund shape as in certain species of wild pigs and bears. The importance of temperature and the evolutionary responses to ambient temperatures should therefore not be underestimated any more than they should be overestimated. Also, temperature is inextricably linked to the water and energy balances of animals even if only in an indirect fashion.

The next questions to pose are, 'How rapidly do adaptations to new temperature regimes evolve and how and when did endothermy evolve?' According to Huey (1982) in a review of temperature, physiology and the ecology of reptiles, the evolution of thermal physiology is conservative for most lizards, with the exception of a few genera such as *Anolis*. He cautions, however, that this view is based on rate of speciation rather than on time. It is also significant that exactly the same neurotransmitters when injected into animals as diverse as giant monitor lizards and sheep elicit similar thermal responses, although in the former they will be mostly of a behavioural nature and in the latter physiological. These results suggest an ancient origin of neurochemical control of thermoregulation which has undergone little change over many millions of years (Bligh *et al.*, 1976). We are, however, not yet in a position to conclude how rapidly adaptations to various temperature regimes evolve. How rapidly, for example, does the ability to synthesise a cryoprotectant or a new suite of isozymes evolve? The acquisition of a thicker fur coat in mammals, however, appears to evolve rapidly as evidenced by woolly-haired feral pigs originally abandoned as short-haired tropical pigs by Captain Cook in New Zealand. The longer-haired, brown hyenas that patrol the cold, coastal regions of the Namib Desert, and leave the hot interior of the desert to the short-haired spotted hyena provide a similar example (Skinner *et al.*, 1984). Obviously, marked changes in body size and shape will evolve more slowly but how rapidly do adjustments in behavioural thermoregulation evolve? One would guess fairly rapidly as they do not necessarily require the acquisition of specialised anatomical structures. These and other similar questions cannot be answered with certainty at present but a large accumulation of field and laboratory data are now available for analysis in conjunction with population studies backed up by molecular genetics. We can therefore expect some significant advances in this area in the near future.

As far as the evolution of endothermy is concerned, several authors have speculated on the selection pressures which brought about endothermy and its associated advantages and disadvantages, e.g. Bennett and Ruben (1979). We have already examined these advantages together with the major changes in morphology and physiology which accompany the acquisition of endothermy, including the fact that endotherms require more than 10 times the amount of food than ectotherms. The question of when endothermy arose, however, remains to be addressed.

The possibility that dinosaurs may have been endotherms was first suggested by Bakker (1972) and has since been widely debated. However, the review by Bennett and Ruben (1986) of the metabolic and thermoregulatory status of therapsids is perhaps of more direct relevance because of the therapsid–mammalian lineage and the fact that therapsids were also a characteristic feature of the fauna during the late Palaeozoic and early Mesozoic (some 200–245 million years ago). Bennett and Ruben conclude that selection pressure for a greater capacity for sustained activity was responsible for the evolution of endothermy among the therapsids. In support of their contention that the therapsids were endothermic, they cite evidence such as the histological structure of the cortical bone, which is very mammal-like, the presence of nasal turbinals and particularly the large number of traits, shared by monotremes and therians, that are associated with metabolic rate and oxygen transport. They point out that modern reptiles have very limited capacities for sustained aerobic activity and become exhausted very rapidly. In contrast, a high capacity for aerobic metabolism and increased activity are closely associated with endothermy in birds, mammals and, when required, in certain flying insects. Moreover, Martin (1980) has pointed out that endothermy in birds and mammals is also associated with the larger brains in these two groups when compared with reptiles, amphibians and fish and he relates this fact speculatively to greater locomotor and metabolic capacity in the endotherms. A careful assessment of brain size in the large fossil reptiles may therefore shed more light on their thermoregulatory status. In the mean time it would seem as if endothermy may have evolved during either the Palaeozoic or early Mesozoic.

We conclude, then, that although temperature has a profound effect on life, it is only one consideration in the complex relationship between animals and their environment. In terms of overall physiological adaptation, it may in fact be a relatively minor factor and the distribution and abundance of animals cannot be predicted on the basis of their thermal physiology alone. Also, many animals can sustain the transient effects of unfavourable temperatures without permanent damage, whereas dehydration and periodic starvation can be fatal. The importance of behavioural adjustments rather than physiological adaptations in overcoming thermal stress would therefore appear to be greater than in the case of nutritional or water stress.

CHAPTER 2

WATER RELATIONS

Water is naturally essential for life but the efficiency with which animals use water varies remarkably. Witness the almost effortless osmoregulation of a gull drinking sea water and the agonising thirst of the shipwrecked human. This chapter will explore this variability and has a similar structure to the preceding chapter in that we shall first discuss the physical aspects, then certain fundamental mechanisms involved in the physiological control of water balance in animals, before moving to actual case studies.

2.1 Physical properties of water

In nearly every introductory biology text, students are informed that water has ideal physical properties for supporting life. This approach is philosophically incorrect because water was present on this planet long before life arose and it would be far more correct to explain that life processes, as we know them today, have evolved largely around the very peculiar properties of water. Edney (1977) has written a useful and lucid review of the importance of these physical properties and this brief description depends heavily on his account.

The molecular structure of water (Fig. 2.1), as one would expect, determines its physical properties (Table 2.1). Stronger intermolecular forces are the reason for its higher melting and boiling points compared with substances with a similar structure. These intermolecular forces are the result of attraction between the two hydrogen atoms of one water molecule and the oxygen atoms of two adjacent molecules. This hydrogen bonding constitutes an energy store of $20 \, \text{kJ mol}^{-1}$. When water is frozen, almost all the water molecules are hydrogen bonded. If the ice melts approximately 15% of these bonds are broken and the breaking of the bonds during melting requires the heat of fusion of ice ($6 \, \text{kJ mol}^{-1}$ of water). The remaining hydrogen bonds must be broken before evaporation of water into vapour occurs and the energy required for this process is much greater ($44 \, \text{kJ mol}^{-1}$ at $25 \, ^\circ\text{C}$ and $40 \, \text{kJ mol}^{-1}$ at $100 \, ^\circ\text{C}$).

As we all know, water bounces readily off a solid surface but it is difficult to compress. For the latter reason it can be employed in hydraulic systems such

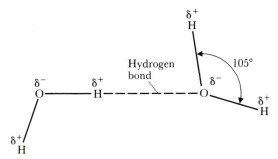

Figure 2.1 Structure of water molecules, showing the hydrogen bonding between hydrogen δ⁺) in one molecule and on an oxygen (δ⁻) in a neighbouring molecule.

Table 2.1 A comparison of the physical properties of water

Liquid	Heat of vaporisation (cal g mole⁻¹)	Melting point (°C)	Boiling point (°C)	Thermal capacity (cal g⁻¹ K⁻¹)	Dielectric constant
Water (HOH)	10 440	0	100	1.000	80.10
Hexane (C_6H_{14})	7 627	−95	49	0.535	1.89
Ethanol (C_2H_5OH)	9 673	−114	79	0.587	24.30
Mercury (Hg)	17 447	−38	357	0.033	—

as the flexible hydrostatic skeletons in many aquatic invertebrates. The flexibility of the skeleton and the poor compressibility of water are exploited in many invertebrate species to provide an appropriate mechanism for locomotion. The high heat of vaporisation of water, and its thermal capacity, are of crucial importance in the evaporative cooling of animals and the distribution of heat in aquatic habitats. The water molecule is polar and water is generally well suited to act as a solvent for the variety of biochemicals and electrolytes which have become incorporated in living organisms during evolution.

Water is not only an important solvent in the tissues of plants and animals, but it also takes part in some of the most fundamental biochemical reactions in living organisms. In plants it is the source of oxygen evolved in photosynthesis and the source of hydrogen for the reduction of carbon dioxide. The high solubility of carbon dioxide in water, and the dissociation of carbonic acid in water, allow the development of the most important biological buffer system, the bicarbonate system. In animals, water takes part in countless hydration reactions when macromolecules are reduced to their simpler components and it is the final product formed in the acceptance by oxygen of electrons in the cytochrome system.

The melting and boiling points of water are such that it occurs as a solid and a gas, but mostly as a liquid at the ambient temperatures which prevail over our planet's surface. Most living organisms have evolved to function within the temperature range at which water is liquid and plants and animals that have to survive outside this range are obliged to synthesise special compounds such as cryoprotectants (see Chapter 1). Because of the unusual hydrogen bonding and van der Waals' forces operating between water molecules in a liquid state, water attains its maximum density at 4 °C. This property allows liquid water to accumulate at the bottom of ponds whereas colder water present at the surface may freeze solid. In this way many forms of aquatic life are able to survive winters in a fluid medium.

Aquatic life also makes use of another important physical property of water, namely its very high surface tension. Small arthropods, possessing an exoskeleton that is strongly hydrophobic may be supported above the surface and will be able to locomote, like gerrids, on the surface of the water. Edney (1977) points out that small hydrophilic insects living beneath the surface of the water, but still dependent on atmospheric air, would experience great difficulty in piercing the water surface were it not for the presence of hydrophobic hairs at strategic positions on their body surfaces (e.g. surrounding the tip of the siphon in mosquito larvae).

We must now consider the vapour phase of water and understand the important difference between relative humidity, water vapour pressure and water activity. Water vapour in the air will exert, like any other gas, a pressure proportional to the concentration of water molecules in that gas system. This is known as the water vapour pressure or partial pressure of the water vapour. The higher the kinetic energy (temperature) of the gas system the more water molecules it is able to accommodate (see Fig. 2.2). Biologists are frequently interested in the 'drying capacity' of air and for this reason calculate the difference between the actual or measured water vapour pressure (P) and the water vapour pressure of the same air at the same temperature when fully saturated with water vapour $(P_s$, see Table 2.2). Thus $P_s - P$ is a useful measure of the potential drying capacity of the air surrounding an animal with a temperature equal to air temperature, and is referred to as the 'saturation deficit'. If the animal's temperature is higher than air temperature, the saturation deficit should not be used as the definitive measurement of the drying potential of air, because the vapour pressure of water on the animal's surface can be higher than the saturated water vapour pressure of the air (Lowry, 1969).

Relative humidity (RH) is not an absolute measure of how much water there is in air. It is the ratio P/P_s, often expressed as a percentage. If we now examine the graphs in Fig. 2.2 more closely it is evident that, although commonly used, RH can be a very misleading measure, when considering the effect of water in air on animals, because the vapour pressure increases curvilinearly with temperature at a fixed RH (see curves A and B). Two environments with the same RH do not necessarily have the same amount of

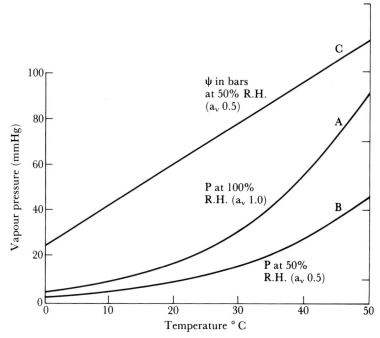

Figure 2.2   The amount of water contained in air at any one relative humidity (RH) or water vapour activity (a_v), increases exponentially with temperature (see curves A and B). Notice how much vapour pressure, and therefore saturation deficit, changes over the temperature range without any change in RH. Redrawn from Edney (1977).

water in the air. Note also from Fig. 2.2. that although the RH differences remain the same between curves A and B over the temperature range (namely 50% RH), the actual vapour pressure differences are small at low temperatures but very large at high temperatures. It is a vapour pressure difference that determines the direction and rate of water vapour movement. Edney (1977) has applied these principles to a practical example showing that water vapour in air on the surface of the desert, even though it is contained in air at a higher temperature and lower RH, will actually move into a cool scorpion burrow where the air is at a lower temperature and higher humidity (Fig. 2.3). The reason becomes clear when one calculates the vapour pressure differences. The same principle holds true for Edney's example of water vapour condensing on the cool surface of a lake (Fig. 2.3).

Considering that it is generally an inappropriate measure of air–water content, it is a pity that biologists (and meteorologists) have become accustomed to using RH, rather than vapour pressure, to describe how much water vapour there is in air. However, there is one circumstance in which RH

Table 2.2 Vapour pressure of water below 100°C. Pressure of aqueous vapour over water in mmHg for temperatures from −15.8 to 100°C. Values for fractional degrees between 50 and 89 were obtained by interpolation. (Reproduced with permission of CRC Press Inc. Boca Raton, Florida)

Temperature (°C)	0.0	0.2	0.4	0.6	0.8
−15	1.436	1.414	1.390	1.368	1.345
−14	1.560	1.534	1.511	1.485	1.460
−13	1.691	1.665	1.637	1.611	1.585
−12	1.834	1.804	1.776	1.748	1.720
−11	1.987	1.955	1.924	1.893	1.863
−10	2.149	2.116	2.084	2.050	2.018
−9	2.326	2.289	2.254	2.219	2.184
−8	2.514	2.475	2.437	2.399	2.362
−7	2.715	2.674	2.633	2.593	2.553
−6	2.931	2.887	2.843	2.800	2.757
−5	3.163	3.115	3.069	3.022	2.976
−4	3.410	3.359	3.309	3.259	3.211
−3	3.673	3.620	3.567	3.514	3.461
−2	3.956	3.898	3.841	3.785	3.730
−1	4.258	4.196	4.135	4.075	4.016
−0	4.579	4.513	4.448	4.385	4.320
0	4.579	4.647	4.715	4.785	4.855
1	4.926	4.998	5.070	5.144	5.219
2	5.294	5.370	5.447	5.525	5.605
3	5.685	5.766	5.848	5.931	6.015
4	6.101	6.187	6.274	6.363	6.453
5	6.543	6.635	6.728	6.822	6.917
6	7.013	7.111	7.209	7.309	7.411
7	7.513	7.617	7.722	7.828	7.936
8	8.045	8.155	8.267	8.380	8.494
9	8.609	8.727	8.845	8.965	9.086
10	9.209	9.333	9.458	9.585	9.714

Temperature (°C)	0.0	0.2	0.4	0.6	0.8
42	61.50	62.14	62.80	63.46	64.12
43	64.80	65.48	66.16	66.86	67.56
44	68.26	68.97	69.69	70.41	71.14
45	71.88	72.62	73.36	74.12	74.88
46	75.65	76.43	77.21	78.00	78.80
47	79.60	80.41	81.23	82.05	82.87
48	83.71	84.56	85.42	86.28	87.14
49	88.02	88.90	89.79	90.69	91.59
50	92.51	93.5	94.4	95.3	96.3
51	97.20	98.2	99.1	100.1	101.1
52	102.09	103.1	104.1	105.1	106.2
53	107.20	108.2	109.3	110.4	111.4
54	112.51	113.6	114.7	115.8	116.9
55	118.04	119.1	120.3	121.5	122.6
56	123.80	125.0	126.2	127.4	128.6
57	129.82	131.0	132.3	133.5	134.7
58	136.08	137.3	138.5	139.9	141.2
59	142.60	143.9	145.2	146.6	148.0
60	149.38	150.7	152.1	153.5	155.0
61	156.43	157.8	159.3	160.8	162.3
62	163.77	165.2	166.8	168.3	169.8
63	171.38	172.9	174.5	176.1	177.7
64	179.31	180.9	182.5	184.2	185.8
65	187.54	189.2	190.9	192.6	194.3
66	196.09	197.8	199.5	201.3	203.1
67	204.96	206.8	208.6	210.5	212.3
68	214.7	216.0	218.0	219.9	221.8
69	223.73	225.7	227.7	229.7	231.7

12	10.518	10.658	10.799	10.941	11.085
13	11.231	11.379	11.528	11.680	11.833
14	11.987	12.144	12.302	12.462	12.624
15	12.788	12.953	13.121	13.290	13.461
16	13.634	13.809	13.987	14.166	14.347
17	14.530	14.715	14.903	15.092	15.284
18	15.477	15.673	15.871	16.071	16.272
19	16.477	16.685	16.894	17.105	17.319
20	17.535	17.753	17.974	18.197	18.422
21	18.650	18.880	19.113	19.349	19.587
22	19.827	20.070	20.316	20.565	20.815
23	21.068	21.324	21.583	21.845	22.110
24	22.377	22.648	22.922	23.198	23.476
25	23.756	24.039	24.326	24.617	24.912
26	25.209	25.509	25.812	26.117	26.426
27	26.739	27.055	27.374	27.696	28.021
28	28.349	28.680	29.015	29.354	29.697
29	30.043	30.392	30.745	31.102	31.461
30	31.824	32.191	32.561	32.934	33.312
31	33.695	34.082	34.471	34.864	35.261
32	35.663	36.068	36.477	36.891	37.308
33	37.729	38.155	38.584	39.018	39.457
34	39.898	40.344	40.796	41.251	41.710
35	41.175	42.644	43.177	43.595	44.078
36	44.563	45.054	45.549	46.050	46.556
37	47.067	47.582	48.102	48.627	49.157
38	49.692	50.231	50.774	51.323	51.879
39	52.442	53.009	53.580	54.156	54.737
40	55.324	55.91	56.51	57.11	57.72
41	58.34	58.96	59.58	60.22	60.86

70	233.7	235.7	237.7	239.7	241.8
71	243.9	246.0	248.2	250.3	252.4
72	254.6	256.8	259.0	261.2	263.4
73	265.7	268.0	270.2	272.6	274.8
74	277.2	279.4	281.8	284.2	286.6
75	289.1	291.5	294.0	296.4	298.8
76	301.4	303.8	306.4	308.9	311.4
77	314.1	316.6	319.2	322.0	324.6
78	327.3	330.0	332.8	335.6	338.2
79	341.0	343.8	346.6	349.4	352.2
80	355.1	358.0	361.0	363.8	366.8
81	369.7	372.6	375.6	378.8	381.8
82	384.9	388.0	391.2	394.4	397.4
83	400.6	403.8	407.0	410.2	413.6
84	416.8	420.2	423.6	426.8	430.2
85	433.6	437.0	440.4	444.0	447.5
86	450.9	454.4	458.0	461.6	465.2
87	468.7	472.4	476.0	479.8	483.4
88	487.1	491.0	494.7	498.5	502.2
89	506.1	510.0	513.9	517.8	521.8
90	525.76	529.77	533.80	537.86	541.95
91	546.05	550.18	554.35	558.53	562.75
92	566.99	571.26	575.55	579.87	584.22
93	588.60	593.00	597.43	601.89	606.38
94	610.90	615.44	620.01	624.61	629.24
95	633.90	638.59	643.30	648.05	652.82
96	657.62	662.45	667.31	672.20	677.12
97	682.07	687.04	692.05	697.10	702.17
98	707.27	712.40	717.56	722.75	727.98
99	733.24	738.53	743.85	749.20	754.58
100	760.00	765.45	770.93	776.44	782.00
101	787.57	793.18	798.82	804.50	810.21

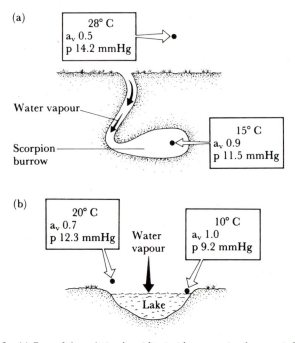

Figure 2.3   (a) Even if the relative humidity inside a scorpion burrow is higher than outside, water vapour will move inwards if the temperature is sufficiently low inside the burrow. (b) In spite of air at the surface of a lake being saturated with water vapour, its vapour pressure may be lower than that of the air above it so water vapour will condense on the lake. Redrawn from Edney (1977).

is the appropriate variable. Some animals and plants take up water from air using hygroscopic salts or concentrated electrolyte solutions. In this case, it is indeed the RH of the air that determines whether the water vapour in it is accessible to the organism.

2.2 Physics of water movement

The movement of water and of solutes in and out of fluid compartments is an exceedingly complex subject but our purpose will be served by reviewing a few key principles.

The rate of diffusion of water can be described by Fick's well-known diffusion equation in its simplest form:

$$\frac{\mathrm{d}s}{\mathrm{d}t} = -DA\frac{\mathrm{d}c}{\mathrm{d}x}$$

where $\mathrm{d}s/\mathrm{d}t$ is the instantaneous rate of movement of substance; D the diffusion coefficient, which is a measure of the ease of diffusion, for example the permeability of the membrane; A the diffusion area; and dc/dx the concentration gradient of the substance moving (moles per unit distance). We conclude that, in a way analogous to heat exchange, the rate of diffusion is largely governed by the diffusion area involved but, instead of temperature gradients, concentration gradients are the driving forces.

Osmosis, the movement of solvent between solutions, is also a complex physical process and opinions differ among physical chemists about the exact mechanisms involved. A highly simplified kinetic model, however, will be used as our example. In this model we assume that the molecules of both solvent and solute are perfectly round spheres and there are two solutions separated by a thin and ideally semi-permeable membrane. Then according to Fig. 2.4A, because of the higher water concentration on the left-hand side, the number of random collisions of the water molecules with the membrane per unit area and per unit time will be greater on the left-hand side than on the right-hand side of the membrane. This difference will result in a net flow of water from left to right.

The mechanism illustrated in Fig. 2.4A does not easily explain the fact that osmotic permeability often exceeds diffusion permeability. Rankin and Davenport (1981) offer a possible explanation for this difference. They suggest that osmotic water flow does not occur by random diffusion as depicted in Fig. 2.4A, but via membrane pores shown in Fig. 2.4B. These pores are water-filled and a pressure gradient is set up within them, down which the water molecules will pass. The motion of molecules along the pore becomes more ordered and less random.

Passive diffusion, which we have been describing, is a simple form of transport during which no energy is released. The process depends on the kinetic energy of the substances involved and takes place down electrochemical gradients until equilibrium is reached on both sides of the membrane. *Active transport* occurs when substances are transported against electrochemical gradients. The exact process is still not fully understood but involves the continual expenditure of energy, mostly derived from the metabolism of ATP. Carrier molecules also appear to play an important role in this process as they can combine reversibly with many substances being transported across membranes.

When discussing actual animal examples we shall emphasise that water itself is never actively transported but on occasion the osmotic gradient is altered in such a way that water appears to move against it. One of the most dramatic examples of this phenomenon is provided by rectal reabsorption of water in insects, which we shall discuss later. As a basis for understanding the possible mechanisms involved in rectal reabsorption, the transport mechanisms across an epithelial cell in Fig. 2.5 should be studied. Active transport of Na^+ leads to a high osmotic concentration within the infolding. Water is then drawn osmotically into the infolding and the increased hydrostatic pressure results in a bulk flow of liquid into the tissue fluid.

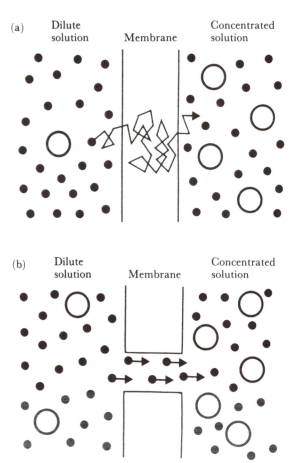

Figure 2.4 (a) A simplified model of passive diffusion of water molecules from a dilute to a concentrated solution (●, water; ○, solute). (b) Uniform flow of water molecules through a membrane pore during osmosis. Redrawn from Rankin and Davenport (1981).

2.3 Terminology

To facilitate the reading of texts and papers, the following elementary terminology should be understood.

Today, the concentration of chemical solutions is almost universally expressed in *molarity*. A 1 *molar* solution contains 1 mole of solute per litre of solution and a 1 *molal* solution contains 1 mole of solute dissolved in 1 kg of solvent. In dilute aqueous solutions the numerical difference between molarity and molality is small but in concentrated solutions it becomes significant. Osmotic concentrations in the tissue fluids of animals are usually expressed

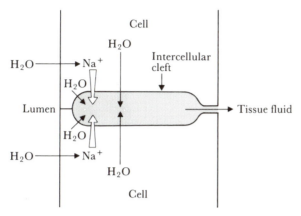

Figure 2.5 A model depicting the transport of water via the intercellular cleft of an epithelial cell. Water follows the active transport of Na^+ and as the volume of the cleft does not change appreciably, water is transported into the tissue fluids. Modified from Schmidt-Nielsen (1983).

in osmoles (1 molal = 1 osmole). Because osmotic activity depends on the number of particles in solution, osmolality depends on the molality of a solution as well as the extent to which the solute dissociates. The osmolality of a NaCl solution, for example, theoretically should be twice the molality, though in practice the dissociation is never complete. Tables are available for calculating the osmolality of electrolyte solutions. The osmolality of sea water is about 1000 mosmol kg^{-1} and that of mammalian plasma about 300 mosmol kg^{-1}.

Osmotic concentrations are usually measured indirectly by measuring another colligative property of the solution, such as depression of vapour pressure or depression of the freezing point. Instruments are available which measure freezing point or vapour pressure depression automatically and provide a direct digital readout in milliosmoles per kilogram.

If two solutions have the same osmotic concentrations they are described as being *iso-osmotic*. If the osmotic concentration of one is lower than a reference solution it is said to be *hypo-osmotic*, whereas the reference solution is now described as being *hyperosmotic* to the former solution. The terms iso-tonic, hypertonic and hypotonic encountered in medical texts are less precise and should not be used.

Some marine invertebrates, particularly those living in estuaries where the osmotic concentration of the external medium fluctuates with the tide, do not regulate the total osmotic concentration of their internal tissue fluids, so that the osmotic concentration of their tissue fluids is the same as that of the external medium; they are called *osmoconformers*. If they are also able to tolerate a wide range of salinities they are described as being *euryhaline*. In contrast, an animal which cannot tolerate a wide range of salinities is known as *stenohaline*. Animals which regulate the osmotic concentration of their body fluids are known as *osmoregulators*. By combining this terminology with that of the chapter on temperature we can come up with some rather overwhelming jargon, e.g. *homeothermic, ectothermic, stenohaline osmoregulator*, which is a reasonable description of a fast swimming tuna fish.

2.4 The measurement of evaporative water loss and water turnover rates

The simplest method of determining water loss from an animal is to weigh the animal over specific time periods and account for all other avenues of weight loss. For example, if the animal does not defecate or urinate, the major measured weight loss would be the result of evaporative water loss (EWL). Admittedly, the animal will also be losing carbon in the form of CO_2, which is usually ignored but can be measured by analysing the gas exchange of the animal or estimated from the metabolic rate. This so-called 'gravimetric' or weighing method generally requires only the simplest apparatus. However, very precise electronic balances are available today which allow the investigator to record the weight of large animals continuously and accurately. They are specially dampened so that movements of the animals are electronically compensated. These balances have been used to record the weight of large mammals standing in direct sunlight under hot conditions and also that of men pedalling exercise bicycles. The continuous loss in weight which is recorded is the result of EWL from sweating and/or thermal panting, provided that defecation and urination are taken into account.

The gravimetric method does not provide a minute-by-minute estimate of water loss from animals, which is a distinct disadvantage to those biologists interested in relating a specific activity of an animal with rates of EWL. To measure EWL instantaneously, use is made of modern electronic sensors which are very sensitive to ambient water vapour. These sensors can be placed downstream of an animal in an enclosed space, and continuous measurements made of the temperature and RH of the excurrent air. These measurements then can be compared with the temperature and RH of the excurrent air when no animal is present in the system, or an additional sensor can be incorporated into the system upstream of the animal enclosure for comparative purposes. By subtracting the water vapour in the incoming stream from that in the excurrent stream, the water vapour lost from the animal can be calculated (Fig. 2.6). The amount of water vapour in the air can be calculated from the temperature and vapour pressure (or RH) using appropriate tables (*Handbook of Physics and Chemistry*) (see Table 2.3).

Both of the described methods for measuring EWL have the very real disadvantage that the animals employed in these experiments must be restrained under artificial laboratory conditions. It is, however, possible to

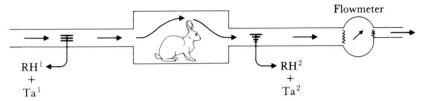

Figure 2.6 By measuring the relative humidity and temperature of the incurrent and excurrent air, as well as the flow rate over the animal, the evaporative water loss of the animal can be measured instantaneously.

Table 2.3 Weight in grams of a cubic metre of saturated aqueous vapour (from Smithsonian Tables). (Reproduced with permission of CRC Press, Boca Raton, Florida)

Temperature °C	0.0	1.0	2.0	3.0	4.0	5.0	6.0	7.0	8.0	9.0
−20	1.074	0.988	0.909	0.836	0.768	0.705	0.646	0.592	0.542	0.496
−10	2.358	2.186	2.026	1.876	1.736	1.605	1.483	1.369	1.264	1.165
−0	4.847	4.523	4.217	3.930	3.660	3.407	3.169	2.946	2.737	2.541
+0	4.847	5.192	5.559	5.947	6.360	6.797	7.260	7.750	8.270	8.819
+10	9.399	10.01	10.66	11.35	12.07	12.83	13.63	14.84	15.37	16.21
+20	17.30	18.34	19.43	20.58	21.78	23.05	24.38	25.78	27.24	28.78
+30	30.38	32.07	33.83	35.68	37.61	39.63	41.75	43.96	46.26	48.67

measure total water turnover rate, though not EWL separately, in the field by using isotope dilution techniques. The most commonly used isotope is tritium (^{3}H). When tritiated water (^{3}HOH) is injected into an animal it soon mixes with the natural water pool of the body and its concentration in the body fluids reaches equilibrium. As time proceeds the isotope disappears from the tissue fluids and if this rate of disappearance or wash-out time is measured, the rate of water turnover in the whole animal can be estimated. In practice, the animal is captured and injected with a known dose of ^{3}HOH. Serial blood samples are then taken over several hours until the isotope has become equilibrated. The animal is then released to behave normally in the field under natural conditions before being recaptured at predetermined intervals for blood sampling and eventual calculation of the water turnover rate (Fig. 2.7).

2.5 Water balance

Whenever one is confronted with the task of evaluating the water relations of an animal, it is advisable to do so by a systematic examination of the water balance, which depicts all the possible avenues of loss and gain (Fig. 2.8). Water balance has been described by Maloiy *et al.* (1979) as an open flow-

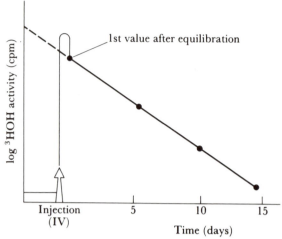

Figure 2.7 Rate of disappearance of tritiated water (^{3}HOH) from the body fluids of a mammal (arbitrary units). The water turnover rate is then calculated from the rate of disappearance of the isotope.

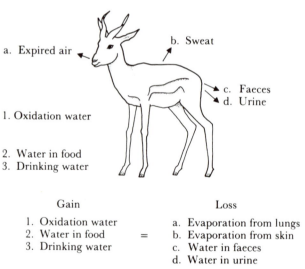

Gain		Loss
1. Oxidation water		a. Evaporation from lungs
2. Water in food	=	b. Evaporation from skin
3. Drinking water		c. Water in faeces
		d. Water in urine

Figure 2.8 Water balance model for a typical mammal, the springbok.

through system. Water enters the system as drinking water, as preformed water in the food and as the water of oxidation (or metabolic water). Water is lost in the urine, faeces and milk and as vapour from the respiratory system and surface of the body. Disturbances in the balance can occur from hour to hour or over several months, depending on the species and the environmental conditions. Negative balances can reach as high as 40% of total body weight in tolerant species, while in some mammals, including man, a loss of 15% can be life threatening. The distribution of water within the various body-fluid compartments also varies, depending upon the degree of dehydration experienced by the animals. For example, in most ungulates dehydration causes a marked reduction in the water contained in the gastro-intestinal tract, the reduction being more severe in sheep and goats than in cattle. The decline in interstitial fluid is less severe but again more marked in sheep than in cattle. In contrast, the water content of the plasma compartment is homeostatically well controlled and undergoes relatively little change during mild dehydration. The opposite situation exists in insects, where the haemolymph (blood) acts as a water reservoir for the tissues.

The type of water balance depicted in Fig. 2.8 is applicable to most mammalian herbivores, but additional avenues of loss and gain must be incorporated if the lower vertebrates and insects are to be considered. In certain birds and reptiles, water loss via salt glands must be considered, while in social insects the phenomenon of trophallaxis, in which members of the same colony feed one another a liquid diet, must be accounted for. In addition, certain insects lose water via repugnatorial glands and when laying pheromone trails. The remarkable ability of certain insects, such as booklice (Psocoptera), to absorb water vapour from unsaturated air, is an additional avenue of gain which must be included.

The major environmental factors influencing water loss in living organisms have been reviewed by Louw and Seely (1982): These are radiation, ambient temperature, the water vapour deficit or saturation deficit and wind speed. All these factors interact with one another and with the inherent behaviour, morphology and physiology of the species in question. In practice, under conditions of high ambient temperature, low vapour pressure, and high wind speed, water loss from animals will be high, especially if the animals exercise. In response to such conditions the animal may be able to increase its intake of free drinking water and reduce water loss via the faeces and urine, or reduce activity.

Before leaving the water balance of animals we should consider the most interesting balance between the production of metabolic water on the one hand and total water loss on the other, because animals that are able to survive without drinking water on a dry diet must produce sufficient metabolic water to equal the amount they lose via all available avenues. Adaptations to ensure such a positive water balance must therefore overcome the problem inherent in the fact that metabolic water production requires oxygen, and delivery of oxygen to the respiratory surface of most terrestrial animals entails water loss via exhaled gases. According to Taylor (1969) large desert antelope such as the oryx overcome this problem by breathing more deeply and not more rapidly. In this way more oxygen is supplied to the alveoli allowing formation of more metabolic water, without greatly increasing the ventilation rate of the dead space in the respiratory system. Taylor (1969) has estimated that under cool night-time conditions the oryx will approach a positive water balance as a result of its increased metabolic rate and slow deep breathing, if the RH of the inspired air does not fall below about 70%. The economics of metabolic water formation is a fascinating problem in ecophysiology but requires far more careful experimentation before exact threshold values can be assigned to different species.

Positive water balances, however, could theoretically ensue from metabolic water production, provided water loss is minimised by suitable adaptations in the following kinds of circumstances, thereby allowing survival without access to free water.

1. Very high metabolic rate plus low respiratory water loss, e.g. flying insect such as carpenter bee.
2. High metabolic rate plus moderate respiratory water loss, e.g. desert rodent in burrow.
3. Very low metabolic rate plus extremely low respiratory water loss, e.g. desert scorpion.
4. Moderate metabolic rate plus moderate respiratory loss, e.g. desert ostrich.
5. Low requirement for metabolic water as a result of large amounts of preformed water in the diet, e.g. gorilla.

The importance of environmental factors such as temperature and vapour pressure in favouring a positive balance are also obvious and well illustrated by Weis-Fogh's (1967) studies at Cambridge on flying locusts. He demonstrated that during level flight a locust would use $270 \, \text{J} \, \text{g}^{-1} \, \text{h}^{-1}$ of metabolic power and, if fat were the sole fuel, it would produce $7.28 \, \text{mg}$ water $\text{g}^{-1} \, \text{h}^{-1}$ of metabolic water. Thoracic ventilation rate in a resting locust is about $30 \, \text{ml}$ air $\text{g}^{-1} \, \text{h}^{-1}$ and rises to $320 \, \text{ml} \, \text{g}^{-1} \text{h}^{-1}$ during flight. Whether or not the locust will remain in a positive water balance during flight will depend on the temperature, radiation load and the RH of the inspired air. Weis-Fogh combined these three variables in a graphic model to describe the ambient conditions which would allow locusts to fly in a positive water balance (Fig. 2.9). He concluded that a locust swarm which was lifted by thermal currents from ground level (T_a, 35 °C) to an altitude of 3 km (23 °C) would be capable of sustained flight in positive water balance if the RH remained above the moderate level of 35%. When the large swarms of these highly destructive insects fly great distances across major desert areas of Africa, they are therefore not likely to be compromised by lack of water.

Weis-Fogh's type of calculation requires more detailed information than is normally available. To estimate the amount of metabolic water production when such information is not available, use can be made of the respiratory quotient (RQ) and estimates of food consumption or energy expenditure. The relationship between substrate oxidised, O_2 consumption, RQ, energy and metabolic water produced is summarised in Table 2.4. Note that

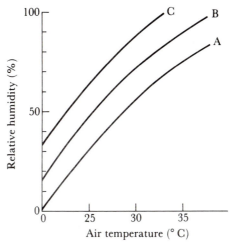

Figure 2.9 Each curve represents a combination of ambient temperatures and humidities at which water loss by a flying locust is balanced by the production of metabolic water. A, no net radiation load; B and C, thoracic temperature theoretically increased by 2 and 4 °C respectively through solar radiation. Redrawn from Weis-Fogh (1967) and Edney (1977).

Table 2.4 Relationship between the respiratory quotient (RQ), metabolic water production and energy production when the major nutrients are oxidised. From Prosser (1973) and Schmidt-Nielsen (1983)

Food	RQ	g water g food^{-1}	$1 O_2$ g food^{-1}	$1 O_2$ g water^{-1}	kJ g food^{-1}	g water kJ^{-1}
Carbohydrates	1.0	0.56	0.83	1.49	17.4	0.0320
Fats	0.71	1.07	2.02	1.89	39.7	0.0269
Proteins	0.79	0.40	0.97	2.44	17.4	0.0228

although fats produce far more energy and metabolic water per gram of substrate than carbohydrates, they actually produce less metabolic water per kilojoule of energy produced. This fact is often overlooked by biologists particularly when speculating on adaptive shifts in the substrate being metabolised to produce more metabolic water.

2.6 Nitrogen excretion

Because protein catabolism occurs in all animals, the problem of excreting the nitrogenous end-products of this catabolism is universal:

$$R-CH-COOH \xrightarrow{\qquad} R-\overset{\overset{\displaystyle O}{\|}}{C}-COOH$$

$$\underset{\text{NH}_2 \qquad \text{deamination} \quad \text{NH}_3}{|}$$

After deamination of the protein the α-ketoacid formed can be fully oxidised to CO_2 and H_2O and excreted through the respiratory and integumentary systems. The NH_3 radical, however, is toxic and is not so easily handled in terrestrial animals. In aquatic animals the NH_3 itself is excreted rapidly into the surrounding water, in which it is highly soluble. These animals are known as *ammonotelic*.

In many animals, including certain aquatic forms, the NH_3 radical is first changed to urea before it is excreted:

$$2NH_3 + CO_2 \rightarrow \underset{\underset{\displaystyle NH_2}{\diagdown}}{\overset{\overset{\displaystyle NH_2}{\diagup}}{C}} = O + H_2O$$

Urea is far less toxic than ammonia and some animals, such as elasmo-branchs, can tolerate very high concentrations in their tissues. They use the raised urea concentration of their tissue fluids, which exceeds that of the surrounding sea water, to allow water to enter their tissues by osmosis. Special adaptations are necessary, however, to be able to tolerate these concentrations. Some animals switch during their lifetime from being ammo-notelic (tadpole) to become ureotelic (adult frog).

The excretion of urea requires the concurrent excretion of a considerable amount of water. Animals that use uric acid as the obligatory end-product for nitrogen excretion enjoy an advantage in dry environments, as uric acid is very insoluble in water and can be excreted in the form of a solid paste or even a dry pellet. The molecular structure of uric acid is more complex than that of urea and requires considerably more energy for its synthesis than urea does:

Uric acid is seldom, if ever, excreted as the acid, but rather in the form of a urate salt and the excretion of uric acid (*uricotelism*) is really urate excretion. Rankin and Davenport (1981) have assembled data on the nitrogenous excretion of Chelonia (tortoises and turtles) to demonstrate how habitat can influence the pattern of obligatory nitrogen excretion even within one order of vertebrates (see Table 2.5).

Table 2.5 Nitrogen excretion in chelonia. From Rankin and Davenport (1981)

Species	Habitat	% of urinary nitrogen		
		Ammonia	Urea	Urate
Chrysemys scripta	Freshwater	79	17	4
Kinixys erosa	Moist terrestrial	6	61	4
Testudo graeca	Dry terrestrial	4	22	52
Gopherus berlandieri	Desert	4	3	93

Table 2.5 cautions us not to be too hasty in assigning nitrogen excretory patterns to different groups of animals. Nevertheless, a broad pattern of excretion does exist:

Aquatic invertebrates	ammonotelic
Insects	uricotelic
Spiders	guanine
Teleost fish	ammonotelic and ureotelic
Elasmobranchs	ureotelic
Amphibians	ammonotelic, ureotelic and uricotelic
Birds and reptiles	uricotelic
Mammals	ureotelic.

Because it is a reasonably small, soluble, molecular particle urea is active osmotically. Though the elasmobranchs can use the osmotic activity to advantage, it would disrupt the osmotic balance of reptilian and avian embryos developing within cleidoic (boxed-in) eggs. Consequently these embryos are uricotelic.

2.7 Excretion mechanisms

The physiology of the various excretory mechanisms in animals such as kidney function, the role of contractile vacuoles, Malphigian tubules in insects and salt glands in reptiles and birds are beyond the scope of this text. For a very readable account of these mechanisms the reader is referred to Rankin and Davenport (1981).

2.8 Case studies

In the case studies that follow, emphasis has been placed on how animals
overcome water stress in hot dry environments because the examination of
function, under the stress of an extreme environment, can frequently reveal
the nature of the mechanisms which are involved.

2.8.1 Marine invertebrates

Most marine invertebrates are stenohaline osmoconformers. Many of them
are able, however, to regulate the ionic composition of their tissue fluids
while remaining iso-osmotic with the external medium. There are also many
species that are exceptions to the generalisation, particularly in estuarine
environments where rapid changes in the osmotic concentration of the external medium can take place with each tide. In such circumstances the
invertebrates survive by becoming either euryhaline osmoconformers or
osmoregulators. An excellent example of a euryhaline osmoregulator is the
shore crab *Carcinus*, which employs nephropores, antennal glands and gills to
regulate the osmotic concentration of its tissue fluids within reasonable
limits. Treherne (1987) has shown that the serpulid worm *Mercierella enigma-
tica* is an extreme form of euryhaline osmoconformer and that its tissues are
adapted to withstand massive changes in the osmotic and ionic concentration
of its body fluids. It is remarkable that the animal's neurones continue to
function normally in spite of the dramatic fluctuation in internal osmocity.
These observations have prompted Treherne (1987) to question the universal validity of the classical generalisation of Claude Bernard in 1865 namely
'The constancy of the internal environment is a condition for free life'.

A modern trend in ecophysiology is to combine genetics, physiology and
biochemistry in an interdisciplinary approach and marine invertebrates have
provided suitable material for these studies. For example, Koehn and
Immerman (1981) and Koehn and Siebenaller (1981) working on aminopep-
tides from mussels (*Mytilus edulis*) showed that allozyme polymorphism was
associated with improved osmoregulation.

2.8.2 Arthropods

Scorpions, ticks and spiders are among the more successful animals in
deserts, mainly because of their resistance to desiccation. In fact scorpions
probably have the lowest evaporative water loss of all animals, an attribute
largely due to the long-chain lipids covering the integument and the sealing
of the respiratory surfaces. Scorpions also have a very low metabolic rate,
which means that the ventilation of the respiratory surfaces is very low. They
can also tolerate high osmotic concentrations of their haemolymph (800 mosmol kg^{-1}) during periods of desiccation and starvation.

Insects represent another major group of successful desert arthropods and their water relations have been well researched. Social insects such as ants and termites are particularly well adapted to survival in the desert, as the division of labour among members of a colony allows them to construct large underground nests in which a favourable microclimate can be maintained. They are also able to store food in the nest and use trophallaxis to transfer watery fluids from one individual to another. The water relations of insects will now be discussed under the headings of waterproofing, minimal excretion, osmoregulation and efficiency of water collection.

Waterproofing

Insect and arachnid waterproofing is largely reliant on lipids upon the epicuticle. Also, arthropods that have low rates of cuticular water loss usually possess more cuticular lipid in the form of hydrocarbons, and long-chain saturated hydrocarbons appear to be associated with superior resistance to desiccation (Hadley, 1977). In addition to cuticular lipids, wax deposits form on the cuticle surface of many insects and arachnids and contribute importantly to waterproofing of such animals. Certain species of tenebrionid beetle, such as *Cryptoglossa verrucosa* from the Sonoran Desert, and at least 26 different species from the Namib Desert produce intricate 'blooms' of waxy threads which are extruded onto the surface of the epicuticle (plate. 2.1). In conditions of low humidity, the amorphous wax secretion becomes organised into many slender filaments (0.14 μm) which interconnect with one another to form a dense network that covers the surface of the epicuticle (Hadley, 1985). This network scatters incident light and the beetle's colour changes dramatically. Why should this waxy bloom develop under conditions of low humidity and what survival value could it possess? The answer to these questions is not entirely clear as yet, but it appears as if the filamentous waxy threads protect a boundary layer of air, with a higher vapour pressure, close to the epicuticle, thereby reducing the vapour pressure gradient immediately above the epicuticle. This boundary layer provides additional protection against desiccation (Hadley, 1979; McClain *et al.*, 1985). For an excellent general review on lipids as water barriers in biological systems, see Hadley (1989).

Evaporative water loss in arthropods, however, does not occur exclusively through the integument. Water loss from the respiratory system is also an important avenue of loss and for this reason xeric arthropods have small spiracles which are frequently hidden or sunken. In some instances they are surrounded by a cluster of outgrowths to reduce loss of water vapour. Hadley (1972) was the first to point out how similar the adaptations in xeric plant stomata are to those found in the spiracles of desert arthropods (see Fig. 2.10). This similarity is not unexpected as they both would benefit from minimising water vapour gradients with the outside air and from the protection of boundary layers around the openings.

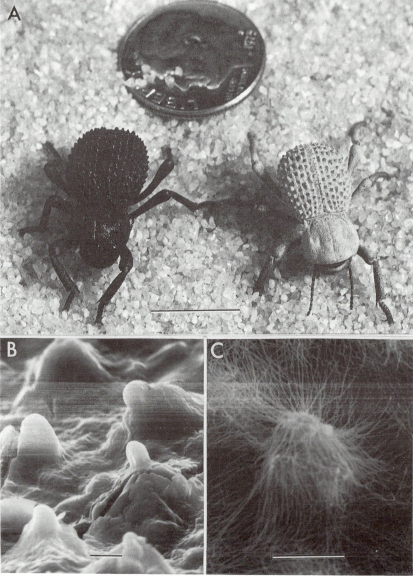

Plate 2.1 When exposed to low humidity the beetle *Cryptoglossa verrucosa* changes to a light blue colour (A). This change is due to the secretion of wax filaments on the surface of the elytra (C) by wax-secreting tubercles (B). The wax filaments protect boundary layers of air on the surface of the beetle, thereby reducing evaporative water loss. Reference bars: A, 1 cm; B and C, 10 µm. (Photo: courtesy N. F. Hadley.)

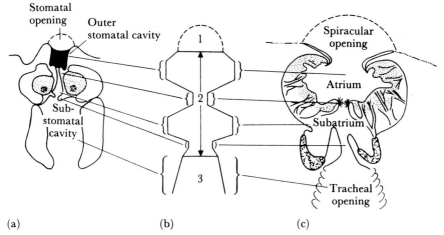

Figure 2.10   Similarities in the structure of a plant stoma (a) and insect spiracle (c) and a comparison of the diffusion resistance in these structures with those in an aperture with varying diameters (b). From Hadley (1972).

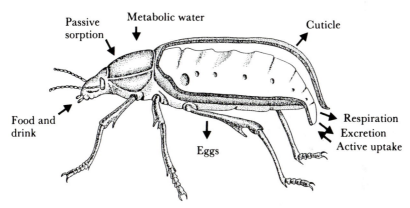

Figure 2.11   The sub-elytral cavity in a flightless tenebrionid beetle, together with major avenues of water loss and gain. Modified from Ahearn (1970).

An advanced adaptation for minimising respiratory water loss in insects is found in the sub-elytral cavity of flightless tenebrionid beetles (Fig. 2.11). The abdominal spiracles open into an air-filled space within the sub-elytral cavity and, because the air space is saturated with water vapour, water loss from the spiracular air is reduced. Expired air leaves the sub-elytral space intermittently via a valve just above the anus (Ahearn, 1970). These flightless beetles also respire discontinuously (Bartholomew *et al.*, 1985), which may afford additional protection against respiratory water losses. The characteristic intermittent respiration persists in one species, *Onymacris plana*, even when running on a treadmill at 35 °C. This beetle sprints over the

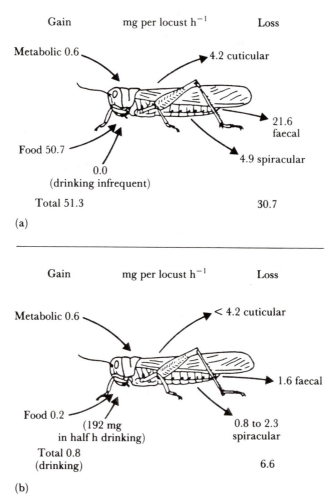

Figure 2.12 The water balance of a locust at 50% RH and 30 °C. (a) Locust feeding on fresh, green grass. (b) Locust feeding on dry food. Avenues of water gain and total gain are shown on the left, and avenues of loss and total loss are shown on the right. Redrawn from Loveridge (1975) and Edney (1977).

surface of hot desert dunes, so it could lose significant amounts of water via respiration were it not for its special adaptations.

Minimal excretion

Arid-adapted insects are ensured minimal excretion of water by the extremely efficient functioning of the rectal pads as well as the Malphigian tubules. This efficiency is well illustrated by comparing the water balance of a locust when feeding on fresh green grass with one feeding on dry food (Fig.

2.12; Loveridge, 1975). When a locust is fed on green grass the most import-ant avenue of gain is the water contained in the food (50.7 mg h^{-1}) and the major avenue of loss is via the faeces (21.6 mg h^{-1}). The insect is therefore in a positive water balance and excess water can be lost either by increasing the ventilation rate or via the faeces. When a locust is fed dry food the balance changes dramatically. Food intake is greatly reduced and only 0.2 mg H_2O h^{-1} is ingested. Similarly, faecal loss is markedly reduced from 21.6 to 1.6 mg h^{-1}. Cuticular loss is slightly reduced, whereas spiracular loss is significantly reduced. Of equal interest is the fact that metabolic water production remains the same at 0.6 mg h^{-1} and the animal is in a markedly negative water balance. In this situation, given the opportunity, a locust will drink as much as 192 mg in 30 min. Free water is frequently absent, however, in their natural habitat so the figures emphasise how dependent locusts are on green foliage with a reasonably high water content (Loveridge, 1975).

Although locusts are usually considered to be reasonably well adapted to desert conditions, the tenebrionid beetles are far superior in this regard (Nicolson, 1990). The tenebrionids of the Namib Desert have been especially well studied and Hadley and Louw (1980) have shown that total water loss from one species (*Onymacris plana*) is as low as 0.1 mg cm^{-2} h^{-1}. This low loss is due to several important adaptations including: a very low metabolic rate; very low respiratory water loss due to protection from the sub-elytral cavity; intermittent respiration with intervals of as long as 60 min between cyclical peaks of O_2 uptake and CO_2 release; sunken thoracic spiracles and water-proofing of the cuticle by long-chain hydrocarbons; and, finally, very efficient reabsorption of water from the faeces by the cryptonephric complex in the rectum (Nicolson, 1980; Louw *et al.*, 1986).

In spite of these remarkable adaptations, many of the desert tenebrionid species still require occasional access to drinking water. The reason for this requirement is that many of these species are obliged to survive on dry food which, combined with a very low metabolic water production, results in a slow but progressive loss of water and reduction in blood volume. The osmotic concentration of the haemolymph is well controlled in the face of desiccation (Fig. 2.13; Nicolson, 1980).

Collecting water

Water collection from the environment is an important way for insects to balance their water budgets. What appears to us as a minute drop of dew can be sufficient to drown a small insect or provide it with sufficient water for many months. We have seen that certain Namib Desert tenebrionids exhibit a very low metabolic rate. The low rate has two important advantages in the desert: it reduces the amount of food energy required and minimises water loss from the respiratory system because of the reduced rate of gas exchange. The major disadvantage, however, is the small amount of metabolic water produced, which obliges the animals to collect free drinking water. Fortu-

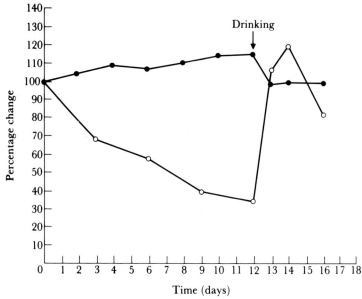

Figure 2.13   Changes in the haemolymph of the desert beetle, *Onymacris plana*, during dehydration and rehydration. Changes in haemolymph volume (o———o) and osmolarity (•———•) shown as percentages of original values. Drinking (day 12) marked by arrow. From Nicolson (1980).

[a]

[b]

[c]

Plate 2.2 Namib Desert dune beetles collect vitally important fog water in different ways. *Onymacris unguicularis* moves slowly to the crest of the dune
(a) where it assumes a head-down posture into the fog wind. Water
droplets condense on the dorsum and run towards the mouth parts
where they are imbibed (b). In contrast, the sympatric *Lepidochora
discoidalis*, a flat pill-shaped beetle constructs trenches to collect fog
water [c]. (Photos: courtesy Mary Seely.)

nately, regular fogs occur in their habitat and fog droplets condense on the surface of the dunes, as well as on the scattered plants and detritus. Some beetle species merely drink the condensed droplets off plants, pebbles or wind-blown dry grass. Others have developed unusual behaviours to facilitate the collection of condensed fog. For example, the tenebrionid *Onymacris unguicularis* engages in what can best be described as fog-basking. These animals are ordinarily diurnal, emerging on a warm sand surface on the dunes during the day and remaining buried in soft slipface sand during the night. When nocturnal fogs occur, however, they will emerge from the sand and climb slowly to the crest of the dune where fog condensation is greatest. Once near the crest they adopt a head-down stance and face into the gentle wind which is advecting the fog. The fog droplets condense on the surface of the elytra and trickle down to the mouth parts where they are imbibed (Plate 2.2). The mean gain in water, expressed as a percentage of body weight after drinking, is 12% with a maximum of 34% (Hamilton and Seely, 1976). Perhaps even more unusual is the trench-building behaviour of a much smaller, disc-shaped Namib tenebrionid *Lepidochora discoidalis*. These beetles are naturally nocturnal and, when advective fog occurs, they construct narrow trenches on the surface of the sand, perpendicular to the fog wind (Plate 2.2). The ridges of these trenches collect more water than the surrounding sand and when the beetles return along a trench, they flatten it while extracting water from the moist sand. The mean water uptake of the beetles resulting from this behaviour is 14% of their original body weight (Seely and Hamilton, 1976). The physical forces involved in moving water from the moist sand, past the mouth parts, into the oral cavity of *Lepidochora* must still be explained. It is probable that the moisture of the sand is above field capacity and strong capillary forces are not necessary to remove water from the sand. If not, capillary-like hairs present on the mandibles of these beetles could assist in this unusual form of water transport.

Some social insects collect and transport water back to the colony where the water droplets are passed from one individual to another, eventually to become part of the 'social stomach'. Honey-bees transport water to the hive on hot days for an additional reason, namely evaporative cooling. An individual bee is able to airlift as much as 52 mg of water to the hive in a single flight, which is equivalent to 65% of its body mass. The energetic cost of normal flight in such a small insect is very high, and when it is water-loaded the cost rises to a remarkably high figure (equivalent to the production of 93 ml $CO_2 g^{-1} h^{-1}$; a human at maximum sprint produces about 1 ml $CO_2 g^{-1} h^{-1}$). Obviously, the ventilation of the tracheal system must increase greatly to supply the increased demand for oxygen and one might well question whether the increased evaporative water loss resulting from this high ventilation rate may not cancel out any benefit from the water being ferried in the crop of the bee. It turns out, however, that even under desert conditions of 30 °C and 5% RH, water-loaded bees produce almost sufficient

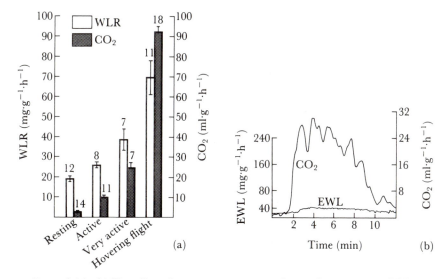

Figure 2.14   (a) The effect of increasing activity on the simultaneous rate of CO_2 production and evaporative water loss in the honey-bee. Vertical bars represent ± 1 SE of the mean; numerals indicate number of replications. (b) Actual chart recording depicting a period of intense activity accompanied by a marked rise in CO_2 production but little change in evaporative water loss (EWL). From Louw and Hadley (1985).

metabolic water $(74 \, mg \, g^{-1} h^{-1})$ to balance the simultaneous evaporative water loss of $79.72 \, mg \, g^{-1} h^{-1}$ during flight. The amount ferried in the crop therefore represents a net gain (Louw and Hadley, 1985). It would appear then that an increased consumption of oxygen is not necessarily accompanied by a proportionate increase in evaporative water loss (see Fig. 2.14). The ability of flying insects to produce as much metabolic water as they lose by evaporation appears to be unique, as birds and bats exhibit the opposite trend (Maddrell, 1987). The major reason for this difference probably lies in the much higher metabolic rate and therefore higher water production in flying insects. The architecture of the tracheole system in insects may also reduce respiratory water losses.

Maddrell (1987) speculatively points out the intriguing possibility that flying insects may employ activity as a means of compensating for water loss. He supports his argument with the fact that Edney (1977) has established that all insects known to be able to absorb water vapour from unsaturated air are wingless. Because they are incapable of the intense activity and oxygen consumption that accompanies flight, natural selection may have diverted them towards evolving alternative water-gaining systems (Maddrell, 1987). A list of arthropods known to take up water vapour from unsaturated air has been compiled by Edney (1977) and it includes Acarini, Thysanura, Mallo-

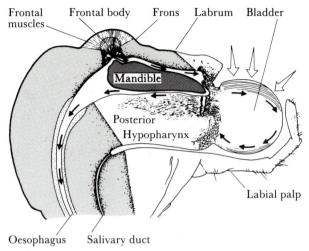

Frontal muscles   Frontal body   Frons   Labrum   Bladder

Mandible

Posterior Hypopharynx

Labial palp

Oesophagus   Salivary duct

Figure 2.15   Head of *Arenivaga* with sections removed to show the distal end of the epipharynx and the hypopharynx. Solid arrows indicate movement of fluid from the site of condensation in the frontal bodies to the oesophagus. Water vapour condenses on the bladders (open arrows) and is drawn into the fluid movements towards the oesophagus. From O'Donnell (1987).

phaga, Psocoptera, Siphonaptera, Coleoptera and Orthoptera. The mechanisms involved in this process are not yet fully understood and we shall limit our discussion to one of the best-studied species, namely *Arenivaga investigata*, the desert cockroach. Absorption of water vapour by *Arenivaga* occurs orally. During absorption two bladder-like structures are extended from the buccal cavity by inflation of the hypopharynx with haemolymph (Fig. 2.15). The cuticular hairs covering these bladder surfaces are strongly hydrophilic and appear to be the most important structures responsible for capturing water at the absorption site. This observation has been confirmed by measuring bladder surface temperatures with micro-thermocouples (O'Donnell, 1987). A non-hygroscopic fluid, apparently an ultrafiltrate of the haemolymph, is produced by the frontal bodies and flows across the bladder surface. Contraction of the frontal muscles forces fluid across a porous cuticular plate and from there to the epipharyngeal groove, and, as RH increases, so does the frequency of muscle contraction. The condensate as well as the fluid applied to the frontal bodies then moves posteriorly and laterally away from the point at the anterior edge of the labrum, where the epipharyngeal groove joins the bladder surface (O'Donnell, 1987). Radioactive tracer solutions that are applied to the bladder surface eventually appear briefly on the posterior hypopharynx and accumulate in the oesophagus. For a detailed and recent discussion of the mechanisms involved in water vapour absorption by arthropods see O'Donnell (1987).

Hygroscopic salt solutions are not applied to the frontal body tissues of *Arenivaga* but salt solutions with an extraordinarily high osmolality (9000 mosmol kg^{-1}) are employed during rectal absorption of water vapour in tenebrionid beetle larvae (Machin, 1981).

Our discussion thus far has been limited to terrestrial arthropods because the problem of an excess of water, which aquatic forms face, is overcome fairly easily. Desiccation, during the drying of temporary ponds and the exposure of arthropods to hypersaline media, however, does present challenging physiological problems to these animals. Anyone who has experienced the transition from a dry to a wet period in a desert can hardly have failed to be impressed by the explosion of life which occurs in a temporary pond. When shallow depressions are flooded by rain after years of apparent lifelessness, bacteria and protozoa appear in great numbers within a matter of hours. Within 1–2 days a variety of small crustaceans appear and grow to maturity rapidly and reproduce as soon as possible. Their large numbers and active feeding habits give the pond the appearance of teeming life. The really interesting question, however, is how the animals survive years of drought between rainfall events without any access to food or water. The answer lies in an extended diapause in the form of eggs which are extraordinarily tolerant of, but not resistant to, desiccation. In fact Clegg (1964) has shown that the desiccation of *Artemia* eggs is a prerequisite for their normal development. Clegg (1987) has also attempted to explain how *Artemia* embryos can survive the loss of almost all intracellular water, a condition known as anhydrobiosis. He concludes that as the encysted embryos of *Artemia* dry out, the cells lose water and become dehydrated, but the damage normally encountered from the loss of primary hydration water is compensated for by its replacement with trehalose. The trehalose, a disaccharide of glucose, protects the structure of the macromolecules and cell membranes but, as desiccation proceeds below 0.3 g H_2O g^{-1} dry mass, metabolic activity ceases. Between 0.3 and 0.65 g H_2O g^{-1} dry mass some metabolism occurs, which is a remarkable reflection of how cell organisation is preserved even in a semi-desiccated state.

Adult, free-swimming *Artemia* are frequently faced with the physiological problem of surviving in hypersaline ponds. They solve this problem by drinking the saline water and transporting the monovalent ions in the gut actively into their tissues. Water follows passively and excess ions are excreted by active transport mechanisms in the gills.

2.8.3 Fish

Marine teleosts
The so-called primitive forms of marine teleosts (e.g. hagfish) are iso-osmotic with sea water but ionic regulation of the tissue fluids still occurs. This

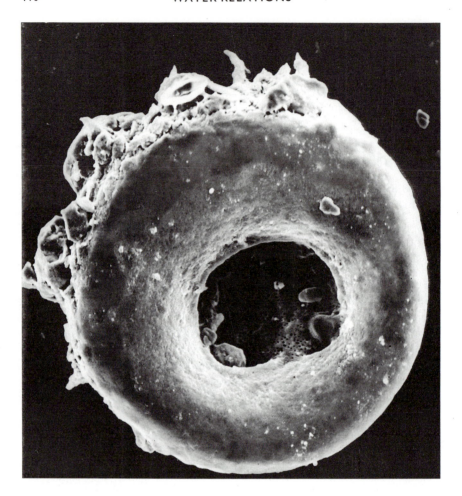

situation is also true for 'old four legs', the famous coelocanth of the Indian
Ocean (*Latimeria chalumnae*). Most marine teleosts, however, are efficient
osmoregulators and maintain the osmotic concentration of their tissue fluids
well below that of the external sea water medium.

Marine teleosts do not possess the loop of Henlé in their kidneys and
consequently cannot produce a hyperosmotic urine. The major difficulty
which they face in sea water is one of dehydration. The gill surface is
specially enlarged to facilitate gas exchange but this adaptation also facilitates the loss of water from the tissue fluids and the influx of ions into the
tissues. Like *Artemia*, the teleosts compensate for this disadvantage by drinking sea water. Active transport of ions then occurs from the gut and water
follows passively. The excess monovalent ions are finally removed by specialised chloride cells in the gills. The divalent ions are excreted directly without
concentration by the kidneys. The glomerulus is consequently greatly

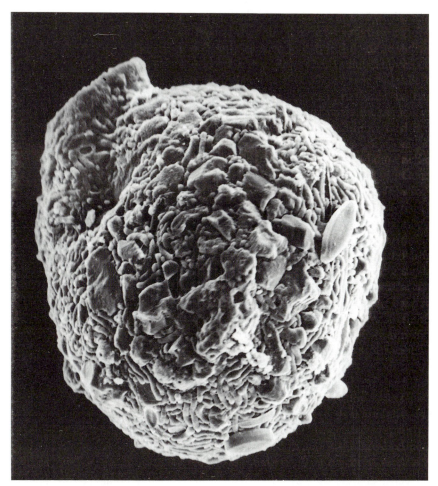

Plate 2.3 Scanning electron micrographs of the remarkable outer shields or tests which protect testate amoeba from periodic desiccation. (Photographs: courtesy C. K. Brain.)

reduced in size so that these teleosts are almost aglomerular. The urine volume is naturally very small.

Freshwater teleosts face difficulties opposite to those encountered in the marine environment: the fish are exposed to the danger of flooding and drowning. One solution to this problem is an engineering one, namely the provision of an efficient pump (the kidney) to remove as much water as possible from the tissues. Consequently urine volumes are large and the urine is very dilute. This mechanism is not sufficient to maintain the required electrolyte concentration in the tissue fluids, so the specialised 'chloride cells' in the gills must function in the opposite direction to those in marine teleosts.

Catadromic and anadromic fish have to cope with the challenge of moving from a freshwater medium to a salt water one (e.g. eel) or vice versa (e.g. salmon). They manage largely by reversing the direction of the ionic pumps in the chloride cells but other adaptations also occur. For example, when eels enter salt water, the epithelium of the oesophagus changes from stratified to simple columnar and loses much of its mucus layer. The question of how the pumps are reversed has not as yet been clarified but it is thought to involve the synthesis of new ATPases which are functionally different (functional variants; see Hochachka and Somero, 1973).

Marine elasmobranchs

These maintain an internal osmolality slightly greater than external osmolality and, as a result, water moves slowly into the tissues from the sea water. Although electrolyte levels are somewhat higher in elasmobranch plasma than in teleost plasma, the high osmolality is largely due to very high concentrations of urea and trimethylamine-N-oxide (TMAO) within the body fluids of these animals. The urea is synthesised from NH_3 obtained from deamination of amino acids and a special segment in the kidney is responsible for urea reabsorption. Elasmobranchs have become so tolerant of high circulating levels of urea that certain enzyme systems and even cardiac muscle will not function normally in its absence.

The high concentrations of urea that elasmobranchs are able to tolerate raises the interesting question of solute compatibility with normal protein function within a cell. Somero (1987) has reviewed this aspect of osmoregulation and concludes that while the solutes NaCl and KCl disturb enzymatic function at high concentrations, the commonly accumulated free amino acids glycine and proline do not have this effect. In contrast, positively charged amino acids such as arginine and lysine do disturb enzyme function strongly. It would seem then that compatible osmolytes or solutes either lack an electric charge, such as the carbohydrates, or carry no net charge as in the zwitterionic amino acids glycine and proline (see Fig. 2.16; Somero, 1987). The potentially disturbing effects of urea on enzyme function in elasmobranchs are adequately compensated for by the stabilising effects of methylamines such as TMAO or glycine betaine. A similar protection exists in the mammalian kidney. It would be interesting to discover how the marine, crab-eating frog *Rana cancrivora* manages in this regard, as it also accumulates large amounts of urea in its tissues.

2.8.4 Amphibians

Amphibians are highly dependent on water for survival and completion of their life cycle. They do not possess a loop of Henlé in their kidneys and are consequently not capable of producing a hyperosmotic urine. The integument of most species is very permeable to water. As larval stages metamor-

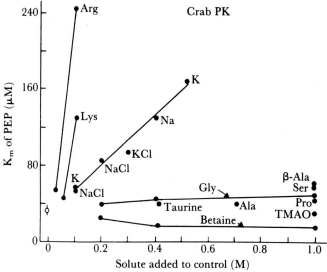

Figure 2.16 The effects of various amino acids and salts on the Michaelis–Menten constant (K_m) of phosphoenolpyruvate (PEP) for pyruvate kinase (PK) of the crab *Pachygrapsus crassipes*. Redrawn from Somero (1987).

phose into the adult form they shift from ammonotelism to ureotelism. Several arid-adapted species, however, have been found to be uricotelic in the adult form and are even capable of storing urates in the bladder. Certain species can reduce the glomerular filtration rate very dramatically, a phenomenon known as glomerular intermittency; urine flow can be completely interrupted, so the phenomenon represents an important water-saving adaptation.

For the reader with the necessary background in physics and mathematics the dynamics of water exchange between a terrestrial amphibian and its environment has been analysed in detail by Tracy (1976). The following examples illustrate some unusual case studies.

Rana cancrivora, the crab-eating frog found in the mangrove swamps of South East Asia, survives in the hypersaline environment of tidal pools by accumulating urea in its tissues.

Phyllomedusa frog species, from South America, have special alveolar glands in the integument which secrete lipids on to the skin surface. The frogs spread the material over the surface of their skins (Fig. 2.17), creating a barrier which reduces the rate of evaporative water loss to as low as that of a typical reptile. Analysis of the lipid secreted by one of the four species (*P. sauvagei*) showed that a complex mixture of triacylglycerols, hydrocarbons, cholesterol and free fatty acids were present but that the dominant product was wax esters (68%) (McClanahan *et al.*, 1978).

Chiromantis, the African tree frog, is abundant in Zimbabwe and it was

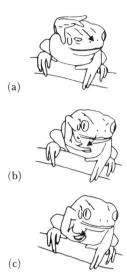

(a)

(b)

(c)

Figure 2.17 Mechanism by which the South American frog *Phyllomedusa sauvagei* employs its forelimbs to spread a lipid secretion over the body surface. Redrawn from Blaylock *et al.* (1976).

Plate 2.4 Certain African tree frogs, such as *Chiromantis xerampelina*, have almost waterproof skins during certain seasons of the year and concentrate the pigment granules in their skins to become very light in colour, thereby increasing their reflectance to solar radiation. (Photo: courtesy J. P. Loveridge.)

there that Loveridge (1970) first showed that frogs could excrete uric acid as the major end-product of obligatory nitrogen excretion. He submitted his findings to the famous journal *Nature*, which rejected them on the grounds of being too revolutionary. Loveridge's results have since then been confirmed frequently. Not only is *Chiromantis* uricotelic, its integument is highly impermeable to water, another very unusual property for an amphibian, and its resistance to desiccation has been compared to that of reptiles (Plate 2.4). This impermeability is probably due to a lipid barrier but the potential barrier has not yet been examined chemically. The eggs of *Chiromantis* spp. develop in a foam nest which is constructed strategically on a branch overhanging a pond or small stream. When the tadpoles emerge they drop into the pond where they complete the larval stage of their life cycle.

Hyperolius species, African reed frogs, frequently bask in the sun and alter the pigment distribution in the integument dramatically for thermoregulatory purposes (Chapter 1). The dorsal skin surface of these frogs is almost waterproof and we have measured EWL rates as low as $1.7 \, \mathrm{mg \, g^{-1} \, h^{-1}}$ in this genus. The ventral surface, however, is not impermeable to water. In fact, it is used to absorb water when the frogs enter streams or small pools. To prevent water loss from the ventrum while basking or resting, the frogs adopt a characteristic posture with the ventral surface and gular region adpressed closely against the substratum and with the feet folded beneath the body (Fig. 2.18).

Of even greater interest, are the recent findings of Shoemaker *et al.* (1987) that two species of waterproof frogs, *Phyllomedusa* and *Chiromantis*, exhibit a marked increase in evaporative water loss at body temperatures of 35 °C and 39 °C respectively. Skin glands are responsible for this sudden increase in fluid loss and the process is analogous to sweating in mammals.

Perhaps even more surprising is that in the desert cicada (*Piceroprocta apache*) active extrusion of water through the cuticle begins at between 39.2 and 39.3 °C. This is the same temperature at which these insects must seek milder microclimates to prevent their body temperatures from reaching lethal limits and, for this reason, this active loss of water is again analogous to sweating and has been implicated in the evaporative cooling of this species (Hadley *et al.*, 1989).

A broad summary of the osmoregulatory responses of adult anurans to aquatic, semi-terrestrial and arboreal habitats has been prepared by Shoemaker (1987). It is most helpful in gaining a rapid understanding of general principles (see Fig. 2.19).

2.8.5 Reptiles

Reptiles, also, do not possess a loop of Henlé and when challenged either by salt loading or desiccation they can reduce urine volume but not increase osmolality beyond that of plasma. They do, however, lay cleidoic eggs and

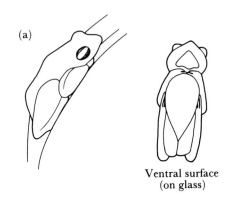

Figure 2.18 (a) Resting position of *Hyperolius*. Ventral surface, gular region and feet are pressed firmly against substratum. (b) Alert position, exposing ventral skin and feet of frog. In this position evaporative water loss increases dramatically. See Withers *et al.* (1982).

AQUATIC

In Water (1 Day)

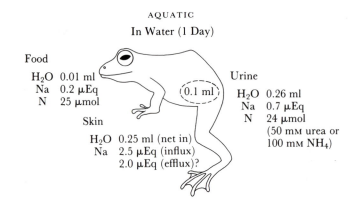

Food

H₂O 0.01 ml
Na   0.2 µEq
N   25 µmol

Skin

H₂O 0.25 ml (net in)
Na   2.5 µEq (influx)
     2.0 µEq (efflux)?

(0.1 ml)

Urine

H₂O 0.26 ml
Na   0.7 µEq
N   24 µmol
(50 mM urea or
100 mM NH₄)

(a)

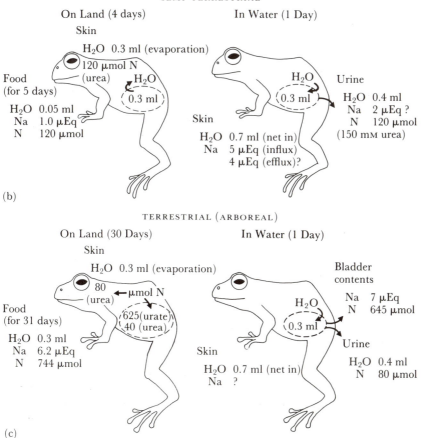

SEMI-TERRESTRIAL

On Land (4 days)   In Water (1 Day)

Skin

H₂O 0.3 ml (evaporation)

120 μmol N
(urea)   H₂O
Food   (0.3 ml)
(for 5 days)

H₂O 0.05 ml
Na   1.0 μEq   Skin
N   120 μmol

H₂O 0.7 ml (net in)
Na   5 μEq (influx)
4 μEq (efflux)?

H₂O
(0.3 ml)   Urine

H₂O 0.4 ml
Na   2 μEq ?
N   120 μmol
(150 mм urea)

(b)

TERRESTRIAL (ARBOREAL)

On Land (30 Days)   In Water (1 Day)

Skin

H₂O 0.3 ml (evaporation)   Bladder
contents

80
←μmol N
(urea)
Food   625(urate)
(for 31 days)   40 (urea)

H₂O 0.3 ml
Na   6.2 μEq   Skin
N   744 μmol

H₂O 0.7 ml (net in)
Na   ?

Na   7 μEq
N   645 μmol

H₂O
(0.3 ml)

Urine

H₂O 0.4 ml
N   80 μmol

(c)

Figure 2.19   (a) Fluxes of water, sodium and nitrogen for an amphibian living and feeding in fresh water (*c.* 1 mEq Na⁺ l⁻¹). All values are per gram of body mass with an empty bladder. (b) Fluxes of water, sodium and nitrogen for a semi-terrestrial amphibian feeding on land (left) and when rehydrating in water (right). Note that a single day in this situation is sufficient for rehydration and excretion of excess solutes. (c) In comparison with (b), the arboreal frogs are well adapted to terrestrial life because of low rates of cutaneous water loss and storage of N-excretory products in the form of urates. This is reflected in the fluxes of water and solute shown. On return to water the bladder is emptied immediately and the frog rehydrates. From Shoemaker (1987).

are uricotelic. Being uricotelic is of great advantage to the desert forms as uric acid can be excreted in the form of a dry pellet with minimal wastage of water. Electrolytes are transported actively from the mixture of urine and faeces in the terminal portion of the digestive tract and cloaca, and water follows the ions by diffusion; a remarkably dry faecal–urinary pellet is produced. The problem remains of what to do with the excess of reabsorbed ions and it is neatly solved by excreting them via nasal salt glands in those species which possess such structures. The uric acid that is voided in reptilian urine is a mixture of acid crystals and minute spheres of urates. The urates also represent an avenue for eliminating excess ions, especially in species which do not have salt glands.

It is popularly thought that the integuments of reptiles are so resistant to evaporative water loss that they are virtually waterproof. Although cutaneous transpiration in many reptiles is lower than in most other terrestrial vertebrates, Dawson et al. (1966) have shown that it is the major avenue of water loss in snakes and lizards, and that cutaneous losses are two or more times greater than respiratory losses. Respiratory losses are particularly low in reptiles, because their relatively low metabolic rates do not require high ventilation rates of their respiratory systems.

In a similar fashion to amphibians, certain species of reptiles store water in a diverticulum of the lower digestive tract which is incorrectly but widely referred to as a bladder. The fluid contained in this bladder is hypo-osmotic to the plasma and can be used as a source of water during periods of water deprivation. Certain tortoises store large amounts of fluid in this way for their long periods of hibernation or aestivation. It is also interesting that within the same genus, *Chamaeleo*, a desert form, *C. namaquensis*, has no sac for storing water but relies on powerful salt glands to maintain water balance under very arid conditions. In contrast, *C. pumilus pumilus*, which inhabits a temperate mediterranean region, relies on storage of water during fairly long dry summers and has no functional salt glands. It would appear then that salt glands are more suitable for overcoming more extreme forms of water shortage.

Obligatory drinking in reptiles is uncommon, and most species are able to balance their water budgets using water contained in their food, supplemented with metabolic water. Some species do drink copiously, however, and in one species by very unusual means. The Australian lizard *Molloch horridus*, in addition to the bizarre spiky projections on its head and body, is covered by a network of fine canals on its integument. When the ventral surface of these animals is placed in water, water moves along these fine canals by capillarity to eventually reach the mouth.

From the point of view of osmoregulation, as well as other physiological functions, reptiles are very well equipped to survival in hot arid regions such as deserts. They have a low (ectothermic) metabolic rate which reduces their requirement for energy, a scarce commodity in the desert, and at the same time this characteristic reduces respiratory water loss. The integument is resistant to evaporative water loss and, because of efficient cloacal reabsorption of water and the excretion of hyperosmotic salt solutions by salt glands, excretory water losses are low. Reptiles are usually small enough to escape from extremes of heat and low humidity. Most are carnivorous or omnivorous, and animal prey contains about 70% moisture, an obvious advantage for balancing water budgets. It is not surprising, therefore, that reptiles are one of the dominant groups of animals in deserts where they have undergone a considerable degree of adaptive radiation.

2.8.6 Birds

Phylogenetically speaking, it is in birds that we first encounter the loop of Henlé. The avian kidney contains a mixture of reptilian-like nephrons or renal tubules and those with the hairpin-like loops of Henlé. These tubules endow them with the capacity to produce a moderately concentrated urine, for example 800 mosmol kg^{-1} in the ostrich. Many birds do not need drinking water, notably insectivorous desert birds and marine birds, but this ability results largely from the efficient reabsorption of water in the cloaca (urodeum) and the lower portion of the intestinal tract, rather than the concentrating capacity of the kidney. Gut water absorption is greatly facilitated by the fact that the obligatory nitrogen excretion of birds is in the form of urates and because the excess of reabsorbed ions can be excreted in a highly concentrated fluid (1000 mosmol kg^{-1}) by the salt gland.

The urine arriving in the urodeum of the cloaca will at best be moderately hyperosmotic to the bird's plasma. It mixes with the faeces in the coprodeum and the mixture is refluxed backwards along the intestine. Ions present in urate salts may at this stage be detached by bacterial action and ions such as Na^+, NH_4^+, Cl^- and phosphate are actively transported into the tissues, with the outward secretion of K^+. Water follows passively and the faecal–urinary mixture can be voided with only a moderate loss of moisture.

The segment of the intestinal tract which is responsible for withdrawing water from the egesta is known as the integrative segment. Amanova (1975) found that in desert-dwelling sparrows the rectal epithelia were well developed and exhibited an increased capacity to absorb water from the luminal fluid, when compared with the epithelia of conspecific sparrows provided with *ad lib.* water. The functioning of the integrative segment is controlled by hormones and so can play a major role in regulating osmotic homeostasis of the body fluids. The field has been well reviewed by Skadhauge (1981) and Thomas (1982). A useful diagrammatic summary of the sequence of physiological events involved in maintaining salt and water balance in birds, together with the major hormones responsible for endocrine control, appears in Fig. 2.20 (Phillips *et al.*, 1985).

Contrary to what one might expect, not all desert birds possess salt glands.

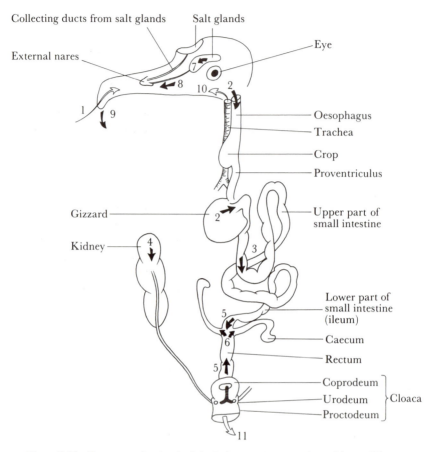

Figure 2.20 Summary of main physiological processes, together with possible controlling factors, involved in salt and water balance in birds. Reproduced with permission of Phillips *et al.* (1985).

Some rather rely on being able to reach distant water-holes by means of their efficient and swift locomotion. One of the best examples is provided by the sandgrouse, *Pterocles* spp. These attractive birds, the size of smallish pigeons, feed in small groups on the arid and semi-arid plains of many parts of Africa. Their major food item is a mixture of very small dry seeds that are mostly overlooked by other animals. The dry nature of their diet and the absence of salt glands compels them to visit water-holes. The precocial sandgrouse chicks hatch from eggs laid on the surface of the desert plain and they start almost immediately to feed on the same dry seeds. They are, however, unable to fly to distant water-holes. The male parent therefore is responsible for ferrying water to the chicks and to do this wades into the shallows of a water-hole to thoroughly wet his breast feathers. These feathers are remarkable in that they can absorb as much as 20–40 ml of water (Cade and MacLean,

Site	Major physiological event	Possible controlling factors
1. Mouth: food and water ingested	Taste Modification of food and water intake (appetite)	Prolactin Angiotensin II
2. Food and water reach upper intestine	Secretions added Beginning of process to achieve isotonicity	
3. Small intestine	Equalisation completed	Thyroxine/T_3
	Gut uptake	Corticosterone Aldosterone Prolactin
4. Kidney	GFR adjusted	Arginine vasotocin (AVT) Corticosterone
	Antidiuresis Isosmotic/ Hyperosmotic urine $\Big\}$	AVT and prolactin AVT aldosterone corticosterone Mesotocin?
5. Lower segment small intestine + ureteral contents mix in rectum	Reabsorption of Na^+ followed by water from iso-osmotic chyme following equalisation process	Prolactin Corticosterone Aldosterone
6. Caeca	Amplification of events in (5).	?
7. Salt gland	Copious production of iso-osmotic fluid	Acetylcholine Corticosterone AVT
8. Collecting ducts	Efficient concentrating mechanism to produce hypertonic exudate	?
9. Beak	Salt gland exudate drips off beak	—
10. Lungs and respiratory tract	Insensible water loss from lungs	—
11. Cloaca	Complementary events to (5).	

Figure 2.20 (continued)

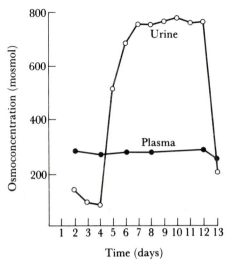

Figure 2.21   Homeostatic control of the osmoconcentration of the plasma of the ostrich during dehydration, largely through adjusted kidney function. The period of dehydration extended from day 4 to day 12. Note that rapid rehydration had very little effect on plasma osmolality. From Louw *et al.* (1969).

1976). The male bird then ferries the water, flying distances of 10–15 km, if not more, to the chicks which then drink copiously from the breast feathers. As much as 3 ml of water has been measured in the crop of a sandgrouse chick, which amounts to 30% of the total body mass of the chick. To compound this remarkable feat, the parent bird also has to locate its small cryptically coloured chicks, which are isolated on a vast monotonous desert plain, and are scarcely detectable by human observers a few metres away (Dixon and Louw, 1978).

Another successful desert bird of Africa and the Middle East, but much larger is the ostrich *Struthio camelus*. Its success in these harsh conditions can be ascribed to various characteristics: it employs body orientation and feather erection to maximise radiant and convective cooling (see Chapter 1); when unable to drink water it reduces the volume of urine dramatically and voids uric acid in a hyperosmotic medium (see Fig. 2.21); it grazes selectively on plants with a high water content; and, when at rest and not engaged in thermal panting, it exhales air that is not saturated with water vapour. The latter phenomenon was discovered by Withers *et al.* (1981) when they placed an electronic humidity sensor in the external nares of ostriches, and made breath-by-breath analyses of temperature and humidity of inspired and expired air. The results appear in Fig. 2.22 and show that the exhaled air from the external nares was not fully saturated. Dropping the expired air humidity from 100% to 85% represents a water saving of 200–500 g day^{-1}. A

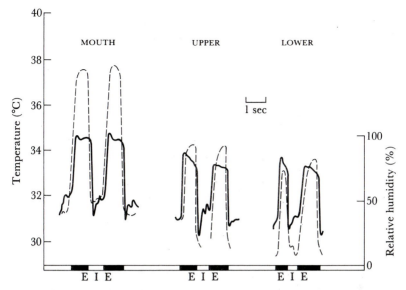

Figure 2.22 Breath-by-breath recorder tracings of relative humidity and temperature of inspired and expired air in the mouth, upper nasal chamber and lower nasal chamber of the ostrich (relative humidity, solid line; temperature, broken line; E, expiration; I, inspiration). From Withers *et al.* (1981).

saving of 500 ml day^{-1} could therefore be of crucial importance to the ostrich, which has a daily water turnover of about 8 l when it has free access to water (Withers, 1983). When compared to the large desert mammals of a similar mass, such as antelope, ostriches have a lower evaporative water loss but, surprisingly, a higher water loss via the urine and faeces.

2.8.7 Mammals

If we return to the original water balance, described in section 2.5, which we used to analyse the water relations of mammals, we could generalise that mammals tend to lose more water through evaporation than most other terrestrial animals but they compensate for this loss by their ability to produce a highly concentrated urine (see Tables 2.6 and 2.7). This statement, however, remains a broad generalisation and certain small rodents and desert goats exhibit remarkably low evaporative water losses.

The ability of mammals, particularly small species, to produce such concentrated urine is usually attributed to the relative length of the loop of Henlé. Anyone, however, who has compared the absolute length of say a large bovine nephron with that of, say, a small desert gerbil under the microscope, may be forgiven for harbouring some doubts about this explana-

Table 2.6 Comparisons of evaporative water loss from various herbivore species. For original references see Louw (1984)

Species	Body mass (g)	Temperature (°C)	Water loss (mg g^{-1} h^{-1})	Water loss* (mg cm^{-2} h^{-1})
Insects				
Desert beetle				
(*Onymacris plana*)	0.7	30	1.34	0.1
Desert beetle				
(*Lepidochora discoidalis*)	0.08	30	2.84	0.1
Locust				
(*Locusta migratoria*)	1.6	30	4.1	0.82
Reptiles				
Desert lizard		field		
(*Dipsosaurus dorsalis*)	50	(summer)	0.35	0.13
Desert tortoise				
(*Gopherus agassizii*)	1.8 × 10^3	23	1.32	1.56
Birds				
Zebra finch				
(*Poephila guttata*)	11.5	23–25	8.58	2.69
Mourning dove				
(*Zenaida macroura*)	118.7	23–25	1.04	0.73
Ostrich				
(*Struthio camelus*)	84 × 10^3	23–25	0.25	1.73
Small mammals				
Kangaroo rat				
(*Dipodomys merriami*)	38.0	28	1.6	0.53
Liomys salvani	55	28	0.6	0.23
Rattus rattus	132.0	28	1.5	0.75
Ungulates				
Bedouin goat (hydrated)	16.3 × 10^3	30	1.08	2.1
Bedouin goat (dehydrated)	16.3 × 10^3	30	0.55	1.07
Sheep	—	38	1.9	—
Grant's gazelle (hydrated)	25 × 10^3	40 day 22 night	1.95	4.45
Grant's gazelle (dehydrated	25 × 10^3	40 day 22 night	1.35	3.08
Oryx (hydrated)	100 × 10^3	40 day 22 night	1.56	5.90
Oryx (dehydrated)	100 × 10^3	40 day 22 night	0.90	3.39
Hereford (hydrated)	225 × 10^3	40 day 22 night	1.95	9.96
Hereford (dehydrated)	225 × 10^3	40 day 22 night	1.60	8.11
Human (sweating)	70 × 10^3	25	—	50.0

* Calculated from original data using the following formulae for surface areas (SA) in cm^2: rodents, $SA = 10 \, (g \, BW^{0.67})$; birds, $SA = 7.46 \, (g \, BW^{0.652})$; ungulates, $SA = 0.142 \, (kg \, BW^{0.635})$.

Table 2.7 Maximum urine concentrating ability in various birds and mammals. For original references see Louw (1984)

Species	Maximum concentration (mosmol kg^{-1} H_2O)	Urine/plasma osmolal ratio
Birds		
Chicken (*Gallus gallus*)	538	1.5
Crested pigeon (*Ocyphaps lophotes*)	655	1.7
Ostrich (*Struthio camelus*)	790	2.7
Zebra finch (*Poephila guttata*)	1005	2.7
Savannah sparrow-salt marsh race	2020	4.5
Small mammals		
Kangaroo rat (*Dipodomys merriami*)	5500	14.0
Hopping mouse (*Notomys alexis*)	9370	25.0
Marsupials		
Red kangaroo (*Megaleia rufa*)	2700	—
Euro (*Macropus robustus*)	2730	—
Bandicoot (*Perameles nasuta*)	4175	—
Ungulates		
Hyrax (*Procavia capensis*)	3088	10.0
Dik-dik	4100	11.0
Bedouin goat	2200	7.0
Grant's gazelle	2789	8.0
Springbok	3000*	10.1
Merino sheep	3200	7.3
Hereford	1160	4.0
Eland	1881	6.0
Zebu	1300	4.0
Camel	3200	8.0

* Theoretical value calculated from kidney dimensions.

tion. A far better correlation exists between urine concentrating ability and mass specific metabolic rate (Greenwald, 1989). It is also difficult to conceive how the urine of *Notomys alexis*, at the remarkable concentration of 9370 mosmol kg^{-1}, does not precipitate proteins and destroy cells as it moves through the renal tissue of this desert mouse (Table 2.7), and how the red blood cells traversing the blood vessels of the renal medulla maintain their integrity.

Faecal water losses in mammals show large interspecific variations. The lowest percentages of faecal water are found among desert rodents: 21% in *Mesocricetus crassus* and 37% in *M. auratus* (Katz, 1973). Arid-adapted ungulates reabsorb large amounts of water in the digestive tract and produce faecal pellets. For example, the dik-dik, German merino, Sinai goat and camel produce faeces with a water content of only 40–50%. In contrast, unadapted ungulates such as cattle, water buffaloes and equids produce unformed faeces with a water content ranging from 70 to 80% (Maloiy *et al.*, 1979).

Adaptations to minimise evaporative water loss in small mammals are largely achieved by appropriate behaviour patterns such as limiting activity to the night and retreating to a closed burrow during the day. The water vapour pressure of the ambient air in a closed burrow will soon rise as a consequence of respiratory water losses from the animal. This high humidity will in turn reduce the vapour pressure gradient between the animal and the ambient air. Another important adaptation found in mammals, and well developed in some small mammals, is the ability to recover water from expired air. The physical process is not completely understood and is usually explained on the basis of countercurrent heat exchange, which involves the cooling of the moist nasal mucosa during inspiration, causing the recondensation of water on the mucosa when the warm, expired air stream returns over the mucosa and is cooled below its dew point (Schmidt-Nielsen, 1972; McFadden, 1983). Although the expired air is still saturated with water, it is at a lower temperature after passing over the nasal mucosa and a net saving occurs. Note that the term countercurrent heat exchange is incorrectly used in these examples; the correct term, recuperative or regenerative heat exchange, has been suggested by Mitchell *et al.* (1987). An even more dramatic adaptation in this regard is exhibited by the camel, which is able to actually dehumidify the expired air in a similar way to the ostrich. In these two instances the expired air is well below saturation level and very significant water savings result (Schmidt-Nielsen *et al.*, 1981).

A reduction in basal metabolic rate is a common adaptation in mammals adapted to arid conditions and some, like the oryx, if forced to increase energy production, and consequently oxygen uptake, do so by breathing more deeply rather than more rapidly (for advantages of this adaptation see section 2.5). The ability of some large herbivores (e.g. camel and oryx) to store heat during the day by allowing their body temperatures to rise and then to lose the stored heat by radiation at night (adaptive hyperthermia) is an important water-saving adaptation (see Chapter 1). Behavioural escape from heat is not limited to small mammals. Large mammals, as we have seen, seek out shade and maximise convective cooling thereby indirectly reducing water loss. If shade is not available the long axis of the body can be orientated to minimise the profile area exposed to direct radiation (see Chapter 1).

Probably still the best-known example of adaptation to arid conditions in small mammalian herbivores is to be found in the pioneering study of the kangaroo rat, *Dipodomys*, by Schmidt-Nielsen and Schmidt-Nielsen (1951). This desert rodent exhibits various traits that are typical of many similar species and which allow survival in arid areas without access to free drinking water. They retreat to underground burrows during the day and forage at night. The entrance to the burrow is sealed and as a result both ambient temperature is relatively low and vapour pressure relatively high, so evaporative water loss is greatly reduced. Over a four-week period, while consuming 100 g of barley, kangaroo rats gained 0.6 ml from water in their food and 54.0 ml of metabolic water. At the same time they lost 13.5 ml of water in the urine, 2.6 ml in the faeces and 43.9 ml by evaporation, giving a total loss and gain of 60 ml (Schmidt-Nielsen, 1964). These animals, like many related species, are therefore able to survive on air-dry seeds through efficient use of metabolic water. Not all small desert mammals, however, are able to survive indefinitely on air-dry food and it appears that this ability is restricted to those that are able to produce urine at a concentration in excess of 4000 mosmol kg^{-1}. The field has been extensively reviewed by Fyhn (1979).

Withers *et al.* (1980), while studying the water turnover rates of small mammals in the Namib Desert, have made an interesting comparison of the ratio of water turnover rate (WTR: ml day^{-1}) to daily energy expenditure (DEE: kJ day^{-1}) among herbivorous mammals. The lowest value of WTR/DEE that is theoretically possible for an animal is determined by the stoichiometry of metabolic water production to O_2 consumption, which in turn depends on the fuel being consumed (Table 2.4). The WTR/DEE ratio for pocket mice was as low as 0.027 ml kJ^{-1}, close to their theoretical minimum. For *Peromyscus collinus*, a Namib Desert rodent, the value was 0.041 ml kJ^{-1}, while among the large ungulates the oryx provided a value of 0.142 ml kJ^{-1} and the non-adapted Hereford steer 0.396 ml kJ^{-1}. It is clear, then, that because of their escape behaviour and remarkably efficient renal concentrating ability, desert rodents are extremely well-adapted herbivores in hot arid regions.

Marsupials

Very extensive areas of Australia experience high temperatures and arid conditions, and in these areas marsupials form an important part of the herbivore community. In view of the variety of body sizes found in marsupials it is difficult to generalise on their adaptations, and our brief discussion will concentrate on the larger species (Macropodidae). Water turnover rates in the red kangaroo and euro are approximately 0.08 ml g^{-1} day^{-1}, and these are among the lower values recorded for large herbivores. Loss of pulmocutaneous water is a major part of the total water loss in marsupials and, as can be expected, it increases rapidly with exercise and in response to rising ambient temperatures. Cutaneous loss has been measured in several marsupials and, although it increases with exercise as expected, it does not do so in proportion to the degree of heat stress experienced by the animal. For example, at 40 °C cutaneous water loss from the red kangaroo and euro amounts to only 1.3 mg H_2O cm^{-2} h^{-1} and 1.5 mg H_2O cm^{-2} h^{-1} respectively (Dawson *et al.*, 1974), whereas in sheep values of from 1.9 to 6.3 mg H_2O cm^{-2} h^{-1} have been recorded (Hofmeyr *et al.*, 1969). Several marsupial species will lick their fur, particularly on their forelimbs, during heat stress, and the blood supply to these areas is increased at high ambient temperatures (Bentley, 1960). Although licking can account for as much as 40% of the total water loss in the bandicoot *Isoodon macrourus* at 40 °C, the importance of this thermoregulatory mechanism has been questioned and varies greatly among marsupial species.

Renal concentrating ability in marsupials is high, and there is no doubt that this contributes to their survival in arid regions. Urine concentrations as high as 4175 mosmol kg^{-1} have been recorded for the bandicoot *Perameles nasuta* and both the red kangaroo and the euro can achieve concentrations of approximately 2700 mosmol kg^{-1} (Bligh, 1973). Faecal water losses in certain species under field conditions can also be reduced to 0.83 g H$_2$O g^{-1} dry weight.

Finally, although all the above adaptations are important in ensuring the survival of marsupials in hot arid areas, they are similar to those found in placental mammals. The peculiar adaptations of the marsupials are to be found in their low metabolic rates and body temperatures, which must be of great significance, when evaluated over long periods, because of the associated reduction in respiratory water loss (Bentley, 1960).

Wild ungulates

A surprisingly large variety of wild ungulates inhabit the semi-arid regions of the world. Their resistance to dehydration varies considerably, and the oryx and various species of gazelle are the most successful in regions of extreme aridity. The oryx, when dehydrated, apparently ceases to sweat and allows its body temperature to rise as high as 45 °C. In view of the animal's high thermal inertia, it is able to accumulate a great deal of heat rather than dissipating it by evaporation, thereby achieving a very significant saving in water, because the excess heat is lost by radiation and convection during the cool night. This adaptive hyperthermia, in conjunction with the oryx's ability to produce a concentrated urine (3100 mosmol kg^{-1}), to reduce faecal water loss and to adjust its respiratory rhythm at night for minimum water loss, allows the species to exist for long periods without access to drinking water (Taylor, 1969). In contrast, most other ungulates such as the zebra, waterbuck, wildebeest, rhinoceros and elephant are obliged to drink regularly.

The small African antelopes, such as the dik-dik and steenbok, are usually confined to wooded areas where, as browsers, they enjoy the advantages of feeding on acacia and other leaves with a high preformed water content. They also seek shade regularly. In the case of the dik-dik, its ability to concentrate urine to 4100 mosmol kg^{-1} and to reduce the water content of its faeces to 44% makes it independent of drinking water (Maloiy *et al.*, 1979). The low thermal inertia of small antelopes limits their capacity for adaptive hyperthermia, which may have favoured the evolution of alternative water-saving mechanisms.

The gazelles are typically open-plains animals and are frequently unable to find shade during the heat of the day. Their urine-concentrating ability is moderately high, and they produce relatively dry faecal pellets. Taylor (1972) maintains, however, that the success of Grant's gazelle in invading the arid regions of Kenya can be ascribed largely to its adaptive hyperthermia. This adaptation is not employed by the springbok, which maintains a relatively uniform body temperature throughout the day and stores heat only when obliged to sprint away from predators (Hofmeyr and Louw, 1987). During these short episodes of heat storage after sprinting, the carotid rete undoubtedly protects the brain from excessively high core temperatures. In addition, as we have seen, the springbok exhibits a well-defined thermoregulatory behaviour pattern to reduce heat gain (Chapter 1). In spite of the importance of these adaptations, our knowledge of how gazelles survive without drinking water, when required to do so, is still incomplete and much further research is needed.

Positive inputs into the water balance are the intake of preformed water and the production of metabolic water. Preformed water in the food of herbivores can range from 5 to 10% in dry seeds to as much as 80–90% in lush pasture and cacti. Many herbivorous animals, including insects and small mammals, will therefore shift their food preferences seasonally to ensure an adequate water intake, even if this means compromising on the intake of certain nutrients and energy. In addition to the differences in water content between plant species, individual plants frequently show a marked diurnal variation in water content (Fig. 2.23). Certain arid-adapted plants, including dormant perennial grasses, absorb water hygroscopically during the night and early morning, when ambient humidity is high, and, if herbivores graze during these periods, a significant increase in water intake can be effected (Taylor, 1968; Broza, 1979; Louw and Seely, 1982).

Domestic ungulates

Certain sheep breeds are surprisingly well adapted to hot, dry conditions, and, given sufficient preformed water in the food, can also be independent of drinking water. Maximum production cannot be maintained, however, in arid conditions. Generally, North European breeds are poorly adapted, while the Awassi, black-headed Persian, Karakul, Namaqua-Afrikaner and Merino breeds are relatively well adapted. Merino sheep possess good urine concentrating ability ($3200\,mosmol\,kg^{-1}$) and produce relatively dry faecal pellets (MacFarlane *et al.*, 1959). The fleece acts as a thermal shield, allowing extensive non-evaporative or sensible heat exchange on the surface of the fleece, thereby reducing the necessity for evaporative cooling. Nevertheless, sheep are not as well adapted as certain goat breeds, and the domestic ungulate best adapted to extreme aridity, apart from the camel, is undoubtedly the bedouin goat. This breed has been studied in considerable detail by Shkolnik and his colleagues (1975) in Israel. They have found that these

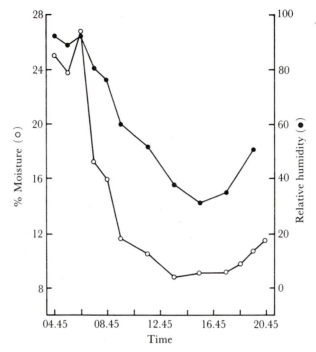

Figure 2.23   The effect of the relative humidity of the air against time upon the
moisture content of a dry, dormant perennial desert grass, *Stipagrostis
uniplumus.* From Louw and Seely (1982).

animals can lose as much as 30% of their initial body mass and then replenish this loss within 2 min when given access to drinking water. Their
black colour gives them a metabolic advantage during winter but is a disadvantage during summer, when they must compensate for the additional heat
gain by evaporating more water than white goats. To be able to sustain this
massive water loss without compromising physiological function is indeed
remarkable and is largely made possible by the capacious rumen. The rumen
allows rapid drinking of large volumes of water when the animal visits
isolated drinking points, which may only be at intervals of 2–4 days. The
rumen apparently also acts as an osmotic barrier, preventing osmotic shock
to the tissues after rapid rehydration (Choshniak and Shkolnik, 1978). Like
camels, bedouin goats continue to feed during dehydration, whereas most
other ungulates will reduce their food intake drastically. This capability,
together with their unusual water metabolism and ability to recycle urea
when on poor pasture, makes them ideal pastoral animals for desert nomads
(Silanikove *et al.*, 1980).

   For example, the implication of only requiring to drink every 2–4 days is
that these goats can range over a much larger surface area in search of food

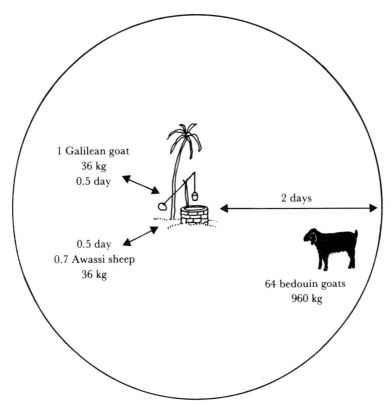

1 Galilean goat
36 kg
0.5 day

2 days

0.5 day
0.7 Awassi sheep
36 kg

64 bedouin goats
960 kg

Figure 2.24 The time interval between drinking determines the foraging distance
from the well. For this reason, and because of their efficient use of the
meagre desert pasture, 64 bedouin goats can be grazed from a single
well point in contrast to one Galilean goat and 0.7 Awassi sheep.
Redrawn from Shkolnik (1977).

than sheep and Galilean goats, which have to drink as frequently as twice per
day (Fig. 2.24). Even when lactating, black bedouin goats drink only once
every 2 days, and their milk production per unit of body mass is the highest
reported for any ruminant.

Other mammals

Humans, the herders of bedouin goats, are in many ways their physiological
opposites. They require highly nutritious refined food, they cannot recycle
urea nor digest cellulose. They have no carotid rete and are unable to store
water. In fact, their evaporative water loss is the highest in the animal
kingdom. Nevertheless, if given suitable clothing (preferably bedouin robes)
and sufficient water to drink they survive very well in hot, arid deserts.
However, even in these circumstances, they do experience the rather minor

problem of a relatively poorly developed water homeostasis. In other words, they tend not to drink sufficient water even when it is freely available. This is also true in polar regions at subzero temperatures. Flavouring of the water makes it more palatable, as shown in the following investigation conducted on young, fit soldiers. The soldiers were divided into four groups and each group was assigned to drink as much as they wished of one of the following fluids: water, water flavoured with fruit extract, milk and beer. The soldiers then took part in vigorous route marches in the Negev Desert. Surprisingly, the group given plain water did not maintain a positive water balance throughout the march. The group given milk developed diarrhoea, which would naturally exacerbate their condition by increasing dehydration. Also, surprisingly, the group on beer were able to maintain their water balance but were not in a suitable condition to engage in modern warfare! The highest intakes were obtained with the fruit-flavoured water and this group maintained water balance far more efficiently, though better replacement fluids have since been devised. Alcohol, because of its diuretic action, is generally not recommended under hot, dry conditions.

In contrast to humans, chacma baboons are able to survive in the dry river beds of the Namib Desert without drinking for periods of 11–26 days. During periods of water deprivation the baboons feed mostly on plants with a high moisture content and reduce all forms of activity to a minimum, including aggressive behaviour among members of the same troop and between troops (Brain, 1991). Carnivorous mammals enjoy the great advantage that their food contains about 70% moisture and consequently many of them are able to survive without drinking water. Their ability to produce a highly concentrated urine is naturally an important asset in this regard and both wild felids and canids are capable of this, although far more research in this field is required. An unusual example of the importance of efficient renal function is provided by the marine fish-eating bat *Pizonyx viveri*, which inhabits certain desert islands in the Gulf of California off the coast of Mexico. Its name is misleading as it feeds mostly on osmoconforming crustaceans with an osmotic concentration close to sea water ($c.$ 1000 mosmol kg^{-1}). It enjoys the advantage of being nocturnal and being able to shelter in the crevices of rocks during the day. The water balance of these animals has been studied in the laboratory with the following results (Gordon, 1982):

Water gain
Amount in food plus metabolic water from food   12.0 ml day^{-1}

Water loss
Evaporation (not active)   3.7 ml day^{-1}
Faecal water   1.3 ml day^{-1}
Urinary loss   6.4 ml day^{-1}
Evaporation (flying)   2.1 ml day^{-1}

Plate 2.5 Chacma baboons are among the most adaptable animals in Africa, feeding in swamps, the inter-tidal zone and surviving even in deserts. These wild baboons have become habituated to the observer and are feeding on termites in the Namib dunes before returning to distant water-holes in a dry river bed. They can survive as long as 26 days without drinking free water. (Photo: courtesy Conrad Brain and Virginia Mauney.)

After only 30 min flying time the bats will enter a negative water balance. Actual flying time has been assessed at 2 hours per night, and on this basis the negative balance would amount to $1.5 \, \text{ml day}^{-1}$. Gordon (1982) reports that the bats drink sea water to make up for this loss and their efficient kidneys get rid of the excess ions by producing a highly concentrated urine. It is instructive also to compare the negative water balance of flying bats with the positive water balance exhibited by flying insects. (Why? See section 2.8.3.)

Marine mammals are a special group in that they are seldom associated with water stress, yet many species feed on osmoconforming marine invertebrates and they are all excluded from sources of fresh water. They do, however, enjoy several advantages in that evaporative losses are kept to a minimum while in or near the water. The air which they inhale is just above the surface of the water and will probably be close to saturation. Nevertheless, as it is warmed in the respiratory system it will take up more water vapour and the animals will lose water via this avenue. Respiratory losses, however, remain small owing to expired air being cooler than core temperature and not being saturated with water vapour (Kasting *et al.*, 1989). Salt loading from the diet is adequately managed by efficient renal function (concentrating ability $c.$ $3000 \, \text{mosmol kg}^{-1}$. Marine mammals also produce a

Table 2.8 Water turnover rates in various herbivore species using ^{3}HOH. Original references in Louw (1984)

Species	Body mass (g)	Temperature (°C)	Water turnover rate (ml g^{-1}day^{-1})
Insects			
Desert beetle(*Eleodes armata*)	0.65	Desert summer	0.053
Reptiles			
Desert iguana (*Dipsosaurus dorsalis*)	50	Desert summer	0.03
Desert tortoise (*Gopherus agassizii*)	613	Desert summer	0.003
Birds			
Zebra finch (*Taeniophygia castanotis*)	13	Simulated desert	0.18
Ostrich—hydrated	95 × 10^3	28–32	0.09
Ostrich—dehydrated	95 × 10^3	28–32	0.02
Small mammals			
Kangaroo rat (*Dipodomys merriami*)	34	29	0.03
Pocket mouse (*Perognathus formosus*)	19	25	0.03
Desert gerbil (*Gerbillus gerbillus*)	34	26	0.13
Marsupials			
Red kangaroo (*Megaleia rufa*)	40 × 10^3	—	0.09
Rock wallaby (*Petrogale inornata*)	3.2 × 10^3	—	0.08
Ungulates			
Bedouin goats	23 × 10^3	30	0.09
Merino sheep	41 × 10^3	25	0.07
German merino	48 × 10^3	15	0.14
Lactating German merino sheep	43 × 10^3	15	0.18
Oryx	136 × 10^3	Winter	0.03
Shorthorn	332 × 10^3	36	0.16
Buffalo	353 × 10^3	36	0.21
Camel	656 × 10^3	25	0.03
Camel	520 × 10^3	37	0.06
Camel lactating	483 × 10^3	—	0.09

highly concentrated milk for their young (40–50% butterfat); compare this to the 3% butterfat in the cows' milk which we drink. This concentrated milk is thought by some to represent an important means of conserving water in these animals. Other schools of thought maintain that the energy-rich milk stimulates rapid growth and deposition of blubber in the young, the latter being an essential tissue for thermoregulation and, in some species, also for buoyancy. The two opinions are not mutually exclusive.

A brief discussion on the scaling of water flux (water turnover rates) in vertebrates is an appropriate topic on which to end this discussion. Louw (1984) has assembled selected data to demonstrate how low the rates of water turnover are in a variety of arid-adapted animals. In all these studies water flux was measured by injecting tritiated water and measuring its wash-

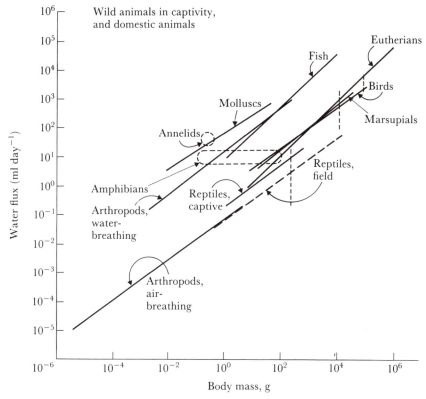

Figure 2.25 Relationship between water flux and body mass in captive wild animals and domestic animals. Axes are logarithmic, and dashed vertical lines are ranges within species. From Nagy and Peterson (1987).

out time. The results are presented in Table 2.8 and show clearly that water flux is reduced very significantly in arid-adapted animals.

Nagy and Peterson (1987) have analysed an exceptionally large data set to study the allometry of water flux in captive and wild animals. Again, all studies involved the use of ^{3}HOH or ^{2}HOH and the results are summarised in Figs 2.25 and 2.26. Nagy and Peterson's allometric analysis show strong correlations among a variety of animals. As expected, water-breathing animals have much higher turnover rates than air-breathing animals, while terrestrial endotherms have higher water fluxes than their corresponding ectotherms. Also, as expected, desert-adapted species exhibit reduced water fluxes when compared to non-desert species. These authors also calculated the ratio of water flux (ml day^{-1}) to energy metabolised (kJ day^{-1}), as measured in the field using doubly labelled water. They term this ratio 'water use effectiveness' (WUE) and it indicates the amount of water used per unit of energy used. As expected (Fig. 2.27), water-breathing animals

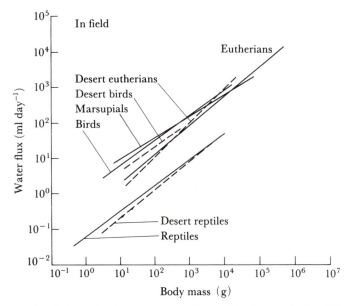

Figure 2.26   Relationship between water flux and body mass in free-living wild animals. Solid lines represent regressions for all species, dashed lines show desert species only (lines for non-desert species not shown). From Nagy and Peterson (1987).

have WUEs more than three orders of magnitude greater than air-breathers.

Of perhaps greater interest is that the WUE calculated for the kangaroo rat under natural field conditions (see ● in Fig. 2.27) is twice as high as when measured in the laboratory while feeding on dry seeds alone. The free-running kangaroo rats may therefore supplement their diet with arthropods or with vegetation with a high moisture content. In any event, they do not have to make continuous use of their remarkable adaptations to minimise water flux. In this way the WUE is a far more accurate reflection of what actually happens under field conditions and prompts the investigator to look more deeply into the natural life style of the animal being studied.

2.9 Concluding remarks

Long before life evolved on this planet, water, with its very peculiar physical characteristics, was universally present. It is therefore not surprising that life, which arose in a fluid medium, has evolved around these peculiar physical properties and that all life forms are still dependent, at least to some extent, on water. In aquatic systems and in temperate terrestrial ecosystems water is abundant and specialised adaptations for its conservation are absent. In arid areas, however, dehydration stress is a reality and it is in

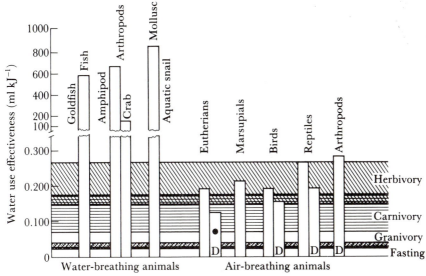

Figure 2.27 Water use effectiveness ratios (WUE) for captive water-breathing animals and free-living air-breathing animals. Filled circle represents kangaroo rat and shaded areas represent WUE values for typical diets, e.g. herbivory (62 and 72% water per fresh mass); animal matter (lipid-rich mealworms and lean whole vertebrates); granivory (seeds with 0 and 10% water per fresh mass); starving (catabolising fat alone). (See text for explanation of WUE.) From Nagy and Peterson (1987).

these regions where one finds the most interesting physiological adaptations. Moreover, it is important to note that, although animals can tolerate transient temperature stress, even short periods of dehydration beyond a certain threshold level will be fatal. For this reason Seely's (1989) view that specialised physiological adaptations to dehydration stress are more important than specialised adaptations to thermal stress in hot arid environments is supported. The latter stress can also be alleviated far more readily by behavioural means.

From the examples discussed in this chapter we can conclude that adaptations to the stress of dehydration on land are numerous and include morphological, physiological and behavioural adjustments. Typical morphological adaptations include the presence of waterproofing material on the body surface, sunken spiracles in insects and colour adaptations to reduce heat loads. Also, specialised structures such as filament-like blooms on insect cuticles that extend vapour pressure gradients, reduce evaporative water loss significantly. Waterproofing is almost universally achieved by the secretion of long-chain hydrocarbons on the surface of the organism and they are employed by animals as different as scorpions, frogs and reptiles. Because temperature is so intimately involved in the rate of evaporative water loss, behavioural adjustments which reduce heat loads will reduce water loss, whereas escaping from microclimates with low water vapour pressures will also be of obvious advantage. Under certain conditions, however, animals must increase evaporative cooling to keep their body temperatures below critical thresholds. Traditionally we associate cutaneous evaporative water loss only with sweating mammals but, as we have noted, this response has now been recorded for other animals as disparate as cicadas (Hadley *et al.*, 1989) and tree frogs (Schmuck *et al.*, 1988). This is an interesting, but not unexpected finding, in view of the universal usage of water by all organisms and its remarkably high heat of vaporisation. Efficient collection of water, e.g. booklice absorbing water vapour from unsaturated air, beetles fog-basking on desert dunes, ungulates grazing during the early hours of the morning or grasshoppers switching their diet to more succulent plants, all contribute significantly to ameliorating dehydration stress.

Physiological adaptations include a wide range of interesting adaptations which either promote the storage of water within the organism, as in the rumen of the bedouin goat and the bladders of certain reptiles, or they minimise water loss via the renal, digestive and respiratory systems. The latter adaptations are well exemplified by the remarkable water-reabsorbing abilities of the mammalian nephron and the insect rectum; and minimising water loss via the respiratory system is well illustrated by the sub-elytral cavity in tenebrionid beetles, the slow deep respiratory rhythm of the oryx antelope at night, and the fact that camels and ostriches appear to be able to expire air that is not fully saturated. Finally, arid-adapted animals generally make efficient use of metabolic or oxidation water within their tissues because various adaptations allow them to balance gains against losses. Of special interest in this regard is the very large amount of oxidation water produced by flying insects during flight, frequently far in excess of the accompanying water loss due to the increased ventilation of the respiratory system. As we have seen, it is only flightless insects which have the remarkable ability to absorb water vapour from unsaturated air. We could indulge in some interesting speculation by saying that, although primitive insect wings may have evolved as heat collectors like some dinosaurs' sails, once they became involved in flight and its attendant requirement for a very high metabolic rate, wings now serve three functions in many insects, namely heat collection, flight and the provision of metabolic water. In other words, a dehydrated insect merely has to spin its flight engines to quench its thirst!

Dehydration stress also occurs in certain aquatic systems. Apart from the extreme examples provided by hypersaline lakes, many vertebrate animals have evolved intricate physiological adaptations to cope with the salinity of normal sea water. These range from urea retention by elasmobranchs through the delicately controlled function of chloride cells in the gills of teleosts to the peculiar breathing patterns of whales. Even the osmoconforming marine invertebrates are able to regulate ionic exchange with their aquatic environment and the survival of *Artemia* in hypersaline lakes as well as its eggs in a virtual anhydric and ametabolic state are, as we have seen, the result of well-controlled physiological adaptations.

When making our concluding remarks for Chapter 1 we stated that gross physiological differences between high- and low-temperature species are not always apparent and that it was seldom possible to provide thermophysiological explanations for the distribution and abundance of animals. Is this also true for physiological adaptations related to water conservation? To some extent yes, but, in view of the more complex nature of the physiological adaptations required to survive dehydration stress, these adaptations probably required a longer period to evolve when compared with fairly simple thermoregulatory behaviour patterns. Moreover, the lethal nature of even transient dehydration stress frequently makes the acquisition of these physiological adaptations essential for survival. Also, it is self-evident how successful and abundant the small uricotelic ectotherms (arthropods and reptiles) are in arid areas and they possess these physiological attributes. Nevertheless, it could be argued that many arthropods and reptiles in temperate areas are also small uricotelic ectotherms. This again emphasises the necessity for a total appraisal of an animal's physiology and behaviour and all the stresses that it is likely to encounter. An alternative solution to extreme aridity is to be a very large animal with a low relative surface area, a high thermal inertia and the ability to locomote efficiently between water-holes in search of scant grazing (e.g. large ungulates). For this reason moderately sized animals are scarce in the desert and are frequently very dependent on underground burrows.

Although the above physiological solutions to survival under extreme aridity are important, if one were pressed to single out the most successful adaptation then perhaps the adoption of a eusocial life should be our first choice. Anyone familiar with areas of extreme aridity will attest to the preponderance of social insects among the local fauna. These are mostly termites and ants but, as we shall see in Chapter 3, also include mole rats and other sub-social species. The advantages of this life style are obvious: the division of labour allows the building of large underground nests with a very favourable microclimate, moist and cool, which can also be used for the storage and growing of food. The division of labour also allows efficient foraging, not only for food, but also for small water droplets which are then transferred (trophallaxis) among nest mates to replenish the water pool of the 'super organism'. In times of profound scarcity the colony can shrink to a minimal size, representing in fact hardly more than the gonads of the super organism. These can then be revitalised with the reappearance of favourable conditions after rain. The evolution of eusociality has, as yet, not been satisfactorily explained although it is thought to have originated in the tropics. Be that as it may, it is in many ways the ideal system for unpredictable, arid environments and aridity seems to be closely associated with eusociality in unusual taxa such as isopods (Linsenmair, 1987), mole rats (Jarvis, 1978) and the recent fascinating finding by Rasa (1990) that a species of tenebrionid beetle in the Kalahari Desert displays social behaviour. In direct contrast, where water is abundant in freshwater and marine ecosystems, there is not a single species of eusocial animal. Although still speculative, aridity may therefore have had a very strong influence on the evolution of eusociality.

To conclude then, physiological adaptations related to the water balance of animals can be used to a limited extent to speculate on the distribution of species. Because of the complexity of these adaptations, they probably have been slow to evolve. Catastrophic climatic change would therefore have a profound effect on animals not possessing these physiological attributes. Nevertheless, the water balance of an animal is closely linked to many other complex morphological and physiological systems such as nutrition, temperature regulation and even reproduction. It is essential then to appraise the total physiology of the organism as well as its behaviour before analysing its ecological status.

What then of the future? Will ecophysiological research continue to emphasise water balances and the osmoregulation of animals? I think not, as many of the gross physiological processes involved in the water balance have been elucidated in a variety of organisms, although we shall continue to be pleasantly surprised by findings such as Hadley's (1989) that cicadas 'sweat' to keep cool. Future research will, I feel, move to the cellular and molecular level to elucidate the structure and function of biochemical control systems and, also, towards population genetic studies to elucidate the heritability and evolution of the physiological traits and to discover whether physiological variation is related to Darwinian fitness. A typical example of the latter type of study is Koehn's (1984) attempt to demonstrate the fitness consequences of allelic variation at loci coding for metabolic enzymes. The role of genetics in physiological ecology has been explored by Koehn (1987) and his views are strongly recommended to the serious student. In addition to these types of studies, recent technological advances in radiotelemetry and isotope studies will facilitate the study of animals moving freely about their natural environment exposed to all the natural stresses but with minimal interference from the observer. Finally, complex and complete biophysical analyses of heat and water balances, such as those carried out by Porter and Gates (1969) and especially the study by Tracy (1976) on an amphibian, are also likely to be attempted in order to test critical hypotheses. The latter will be a natural development as increasing numbers of physicists are cooperating with numerate biologists.

CHAPTER 3

NUTRITION AND ENERGY

When in everyday life humans refer to the well-known phrase 'The struggle for existence' it is interpreted as meaning the procurement of food and shelter. Similarly, when most people think and talk about animals in this context they are frequently conceived to be engaged in a titanic struggle for survival and this struggle centres largely on being able to obtain sufficient food and energy from the environment. Clearly, food or energy are of primary importance in the lives of all animals, but many modern ecologists would caution against the preceding assumptions. They believe that in many natural ecosystems there is more than sufficient food available for a variety of different animals. It has also been questioned if interspecific and/or intraspecific competition for food is a limiting factor in all ecosystems. These questions may appear to be simple but they are of profound importance in ecology and evolution. Moreover, they are very difficult to examine experimentally because of the complexities involved in a natural ecosystem and because it is difficult and expensive to manipulate abiotic and biotic factors on a large scale under natural conditions. We should, however, keep these questions in mind, particularly when examining the energy requirements of animals, because it is very easy to fall into the intellectual trap of believing that evolution has always shaped the lives of animals to expend a minimum amount of energy. In fact, evolution often produces compromises.

In this chapter we assume an elementary knowledge of nutrition and digestion, as well as the many fascinating mechanisms that have evolved for feeding. We shall touch briefly on certain principles involved in these phenomena but the reader who requires more details is referred to a standard physiology text (e.g. Schmidt-Nielsen, 1983 or Eckert and Randall, 1983).

3.1 Nutrition

3.1.1 Nutritional requirements

In spite of the great importance of nutrition, the complete nutritional requirements are only known for a remarkably small number of living organisms. These include a variety of plants, including bacteria (especially

Escherichia coli), some fungi, various insects, the laboratory mouse and rat and domestic animals. Humans have also been well studied but our knowledge is still not complete for the obvious reason that ethical considerations limit experimentation on humans.

In spite of the very few animals that have been studied, certain broad generalisations have emerged, namely that a great variety of mechanisms exists for the acquisition or prehension of food. This diversity diminishes when we examine the various modes of digestion, although considerable variation still exists in this regard. Of greater interest, however, is that very little difference exists among animals in terms of the ultimate requirements of the cells and the ways in which the nutrients are used and metabolised. Consequently, in spite of great differences in their size, life style and food, we would expect elephants and mosquitoes to have very similar nutrient requirements at the cellular level. These would include water, oxygen, essential amino acids, essential fatty acids, essential macro- and micro-elements and the necessary fat-soluble and water-soluble vitamins. In addition, the mosquito would require a source of dietary cholesterol.

3.1.2 Determining nutrient requirements

An understanding of the techniques involved in a discipline often facilitates and enhances the understanding of that discipline and for this reason we shall briefly examine the more important techniques used for determining nutrient requirements.

The *purified diet technique* is used to determine both qualitative and quantitative requirements. It consists essentially of assembling chemically pure nutrients in a diet which theoretically will supply all the nutrients required by the animal in question. Typical items used in such diets are: chemically pure casein, pure hydrogenated oils, chemically pure fatty acids, mineral salts and vitamins. The diet is then fed to the experimental animals and, by the omission and alternative replacement of individual nutrients, essential nutrients can be identified on the basis of the development of deficiency symptoms. The use of radioactively labelled nutrients, supported with blood and tissue analyses, has greatly enhanced the effectiveness of this technique. However, it remains a tedious, expensive and time-consuming procedure and it is not surprising that the complete nutrient requirements are only known for a few animals.

Quantitative nutritive requirements are more usually examined in *growth experiments* and *balance trials*. In growth experiments the nutrient in question is fed at increasing quantities until no further increment in growth of the experimental animals occurs. To study nutrient requirements really accurately, however, they should be studied over long periods on the basis of total intake and total loss over that period—this would constitute a balance trial. For example, nitrogen balances have been frequently performed to study the protein requirements of animals. In these trials various amounts of the protein in question are fed to the experimental animals and the intake of N in the food is measured daily. This is then compared with the daily losses of N in the urine, sweat, feathers, hair and faeces. The daily retention and ultimately the optimum daily intake can then be calculated. These trials again represent rather tedious research but they are of great practical importance.

3.1.3 Determining the nutrient content of foods

Ecologists and animal nutritionists often require a rapid estimate of the nutritive value of a food. In these cases use is usually made of the so-called *proximate analysis of a food*. This analysis consists of determining the total N in the food and then multiplying the amount obtained by 6.25 to obtain the value for crude protein, because most proteins contain 16% N. The food is then subjected to ether extraction, boiling with alkalis and acids and finally ashing in a muffle furnace. In this way the fat content is estimated from the ether extract. Carbohydrates are determined by difference, i.e. the amounts of all the other nutrients are added together and subtracted from the total dry weight of the original sample. For this reason this component is usually known as the nitrogen-free extract (NFE). Finally the cellulose or crude fibre is determined by removing all nutrients from the sample except the crude fibre and the minerals by boiling the sample first in an alkali and then in an acid. The remaining slurry is dried, weighed and the crude fibre is burnt off in an ashing oven. The remaining ash is weighed and the amount of crude fibre is determined by difference.

This is a useful technique for rapid results but it has many inherent errors. For example, the assumption that all proteins contain 16% nitrogen is not correct, nor does ether extraction only remove fats and oils, it also removes plant pigments, some vitamins, steroids and other fat-soluble items. These objections are, however, minor when compared with the absence of any digestibility data. A nutrient may be present in high concentrations, but if it is not digestible it is useless to an animal. For this reason a more refined version of the proximate analysis has been developed. It is very similar in that the same chemical methods are used but known amounts of the food are fed to the animal being studied while simultaneously collecting the faeces voided from the experimental food being fed. The faeces are then dried, weighed and subjected to the same proximate analysis as the food. In this way the percentage digestibility of each nutrient is determined:

$$\frac{\text{Crude protein in food} - \text{Crude protein in faeces}}{\text{Crude protein in food}} \times 100 = \begin{array}{l}\text{Digestibility of} \\ \text{crude protein in} \\ \text{food}\end{array}$$

Once all the digestibilities have been determined, they are applied to the original amounts in the food as revealed by the proximate analysis of the food. The amount of crude fat is multiplied by 2.25 as fat contains 2.25 times the amount of energy contained in the other nutrients. The result of each of the calculations is added together to provide a far more useful index known as the *total digestible nutrients (TDN)*.

The *energy content* of foods can be similarly refined, in fact more so. If we merely determine the gross energy in a food by burning it in a bomb calorimeter, we have a very crude measure. For example, straw and oatmeal porridge would give similar results, both possessing an energy of about $14.7 \, kJ \, g^{-1}$. If, however, you were presented with these two items for breakfast, you would unhesitatingly avoid the straw because of its low digestibility and high cellulose content. The amount of digestible energy in a food is therefore a far more useful measure and energy evaluations can be progressively refined as follows:

Total heat produced in
bomb calorimeter   = Gross energy (GE)
Digestible energy (DE)   = GE − energy in faeces (FE)
Metabolisable energy (ME) = GE − (FE + energy in urine (UE))*
Net energy (NE)   = GE − (FE + UE + energy in voided gases, e.g. CH_4 + energy lost in heat increment of feeding†)

In spite of the usefulness of the proximate analysis and total digestible nutrient techniques, they are seldom used by ecologists. Metabolisable energy and protein content determinations are the most commonly used methods at present.

The *quality of proteins* present in a foodstuff is an important concept and is not revealed by crude measurements such as the proximate analysis or even total digestible nutrients. For example, if one were to feed laboratory rats on a diet containing 20% of a low quality protein (e.g. keratin—the protein in horns, nails and hair) and were then to compare their growth with a group on 10% egg yolk, the latter group would grow far more vigorously and be in better health. Protein quality is therefore determined not only by the digestibility of the protein but primarily by the amount and ratio of the so-called essential amino acids in the diet. The requirements for essential amino acids differ from species to species but are remarkably similar for most animals studied thus far, e.g. requirements for albino rats are phenylalanine, histidine, isoleucine, leucine, lysine, methionine, threonine, valine, arginine and tryptophane (PHILL MT VAT). Protein quality is therefore of critical importance in the lives of most animals. Certain animals are, however, capable of existing on diets with a very poor quality of protein, e.g. certain insects and ruminants. These animals possess micro-organisms in their digestive tracts

---

* With ruminants the energy value of methane gas produced during digestion must also be subtracted.
† Heat produced after consuming a meal; to be discussed later in some detail.

which can synthesise amino acids for their own use from organic nitrogenous compounds. In fact in the case of ruminants, a very high quality protein will actually be degraded somewhat in quality as it is transformed into bacterial proteins. Ecologists must be acutely aware of these differences when assessing the partitioning of resources within an animal community. Protein quality is determined somewhat crudely in growth and digestion experiments and accurately in balance experiments.

Before leaving nutritional experiments, first a warning about relying too heavily on live weight as a criterion for growth. The mere mass of an animal gives no indication of the chemical composition of its body and although one diet may produce far swifter live-weight growth than another, it may produce a very lean body with low fat (energy) content. For example, we may carry out an experiment by feeding different food plants to the larvae of an important insect. If we first kill a representative sample at the beginning of the experiment and then again from each experimental group at the end of the experiment, we could analyse them chemically for fats, proteins, carbohydrates and minerals or merely determine their total energy in a bomb calorimeter. This procedure would provide us with far more meaningful information than merely weighing the animals. It is of course not always possible to kill animals for this purpose, particularly rare animals and large vertebrates. In these circumstances the amount of water can be determined in a live animal by injecting it with tritiated water (^3HOH) and then calculating the dilution of the isotope. Once the amount of water in the animal is known the amount of fat can be estimated from established regression equations which predict the very strong negative correlation between fat and water in the vertebrate body. Finally the ratio of protein to minerals, the remaining major chemicals, is very constant in vertebrates on a fat-free, water-free body mass basis. In this way a surprisingly useful estimate of *body composition* can be made in terms of water, fat, protein and minerals if suitable regression equations exist for the same or similar species (Fig. 3.1).

3.1.4 Food chains

Ecologists go to great pains, and rightly so, to construct food chains and food webs within the ecosystems they study. The results are usually very informative and have provided us with important elementary principles, such as the 10% law which states that about 90% of the energy in one trophic level is lost as respiration when being transformed into the energy of the next trophic level. This means that only a 10% net usage is made of the energy from each trophic level as we proceed up the food chain from primary producers (plants) through primary consumers (herbivores) to secondary consumers (carnivores) to tertiary consumers (top carnivores). The 10% law also explains why food chains are relatively short. In an excellent popular book by the eminent ecologist Paul Colinvaux, *Why Big Fierce Animals Are Rare*, the

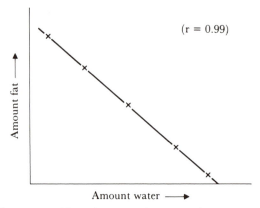

Figure 3.1   The amount of fat and water in the tissues of most vertebrates shows a strong negative correlation.

explanation of the title is obviously provided by the 10% law, i.e. the steep attenuation of available energy as one proceeds up the food chain—there is just not sufficient energy available at the higher levels to support large populations. Here are some unusual food chains merely for interest's sake.

Along the south-western coast of Africa strong upwelling of cold nutrient-rich water occurs in the Atlantic Ocean. When these nutrient-rich waters reach the surface and travel past the desolate and dry coast of Namibia, the populations of phytoplankton and zooplankton explode and huge shoals of sardines and other small fish feed heavily on the plankton. The fish in turn are consumed by marine birds and fur seals that roost on the desert coast. Normally, this is where the food chain would end but the dissection of the desert by dry river beds allows lions to move in small numbers from the fairly temperate, inland escarpment through the desert to reach the coast. Here, one can be greeted with the extraordinary sight of lions feeding on Cape fur seals.

Another illustration of the unusual directions food chains can take is provided by algae growing on the underside of the ice floes in the Antarctic. Small crustaceans cling to the underside of the ice and feed upon the algae. Many of these krill-like creatures eventually coalesce into huge schools and are consumed in large quantities by baleen whales. The whales in turn are harvested and some of their oils and fats are used to manufacture cosmetics such as lipstick. Imagine the path that the lipids have taken from the trapping of sunlight by algae beneath the ice in the freezing waters of the Antarctic to the lips of, say, a wealthy matron in Phoenix, Arizona.

To conclude this brief section on food chains, it is important to note that the extent to which the primary plant production is consumed by herbivores differs markedly among various biomes and ecosystems:

	Percentage consumed
Phytoplankton	60–90
Grasslands	12–45
Kelp beds	10
Spartian marshes	7
Mangroves	5
Deciduous forests	1.5–5

This brings us to the interesting subject of palatability.

3.1.5 Palatability and the chemical defence of plants against herbivory

Palatability is such a common and everyday phenomenon in our lives that few of us stop to consider the physiological implications of taste and palatability. The human response to the four basic taste stimuli, sweet, sour, salt and bitter, is well known but organic chemists working in the food and wine industry today recognise far more complex and subtle stimuli which involve the olfactory system as well. They have also been successful in isolating and synthesising many of the flavours that humans prize, notable exceptions still being coffee and chocolate. In spite of the highly developed ability of humans to recognise tastes and odours (for example, some wine connoisseurs can recognise variety, vintage and even the name of the vineyard where a wine was produced), humans nevertheless have poorly developed homeostasis for eating a balanced diet. Cafeteria experiments have shown that the major natural hunger drive in humans is for calories with sugar or sweetness representing a super stimulus. In addition, a strong craving for salt develops when Na^+ is removed from the diet. It is not surprising then that peace treaties in ancient times often included the exchange of large quantities of salt but perhaps it is only an amusing coincidence that the recent strategic arms limitation treaty between the USA and the USSR was known by the acronym SALT. Marooned sailors and refugees forced to eat a monotonous diet of plant products report very strong cravings for meat and animal fats. These cravings do not, however, add up to good nutritional homeostasis and many urbanised humans suffer from gross malnutrition by living on white bread, soda pop and french fries. In London the susceptibility of middle-aged single women to develop malnutrition was designated as the 'tea and toast syndrome'. In contrast to humans, if the laboratory rat is provided with a cafeteria diet, it will in most instances eat a fairly balanced diet and remain healthy.

In spite of our extensive knowledge about palatability and taste in humans, very little is known about this phenomenon in most animal groups. For example, humans can be fooled into accepting artificial sweeteners, such as saccharin, but the honey-bee immediately rejects these compounds and remains faithful to pure natural sugar. How does sugar elicit a neuronal response in humans? Is it the result of an interaction between the sugar molecule and the interface of a receptor site on or near the neurone, which can be mimicked by the molecule in the artificial sweetener? Is the receptor site in bees slightly different, preventing the artificial sweetener from eliciting a response? These are all interesting questions but no answers are available as yet. Many more examples of our ignorance about the palatability of the natural foods of animals could be cited. How are dung beetles attracted over long distances to freshly voided herbivore dung? Why does alfalfa have almost a universal appeal to herbivores? We do not know, and our present knowledge is confined to a limited number of insects, of which the silkworm has probably been the best studied. The silkworm, *Bombyx mori*, feeds almost exclusively on the leaves of black and white mulberry trees. The substances that provide the feeding stimulus for these larvae are divided into three groups: olfactory attractants, biting factors and swallowing factors. The larvae are first attracted by an olfactory stimulus due to a mixture of mono-terpenes in the mulberry leaves such as the essential oils: citral, terpinyl acetate, linalyl acetate, linolol, etc. Once the larvae have been attracted to the leaf they are further stimulated by biting factors and particularly by the flavonoids, morin and isoquercitrin. Common sugars such as sucrose and inositol also stimulate the biting response while cellulose, phosphates and silicates are involved in the swallowing response (Hamamura *et al.*, 1962; Harborne, 1982). Sufficient knowledge on feeding stimuli in insects has accumulated to allow entomologists to make up a variety of artificial diets for raising many insect species under artificial conditions (Singh and Moore, 1985).

To be highly palatable is naturally not always to a plant's advantage and for this reason many plants have evolved *chemical defences* against being eaten. Some of these defences are permanent and others are temporary. For example, when a fruit is still green and the seed within has not reached maturity, the fruit frequently has a high tannin content and is most unpala-table. Once it has ripened it is usually advantageous to the plant for the seed to be dispersed and consequently the tannin content is reduced, sugar levels rise and the fruit turns an attractive bright colour. In most instances, how-ever, these so-called *secondary compounds* are a more permanent feature of plants' chemical defence and they are particularly prominent in plants exper-iencing some form of stress. Coniferous forests in cold regions grow very slowly and depend heavily on secondary compounds to defend their needle-like leaves against herbivores. Also, plants growing on shallow mountain soils of low fertility often grow slowly and retain their leaves for long periods. Similarly, desert plants and even tropical plants growing in unfertile sandy soils employ some form of chemical defence to deter herbivory. The most commonly employed compounds are the *tannin polyphenols* that impart a highly astringent taste to the tissues of these plants. When the leaves are finally shed and decompose, the liberated tannins in the soil colour the water in neighbouring streams and ponds a brown tea-like colour. Anyone who is not certain of the meaning of astringency need only bite into a green olive or persimmon. In spite of the widely known deterrent effect of tannins, they do not always deter herbivory and their presence in plants should be interpreted with caution. For example, Raubenheimer and Simpson (1990) showed that the presence of tannins in the diet of the locust *Schistocerca gregaria* actually increased food intake in both choice and no-choice experiments. This effect resulted in these insects being heavier at adult emergence. The study of secondary plant compounds has become a discipline in itself and the following brief remarks are merely to introduce the reader to the subject. For more details the student is referred to a very readable review by Harborne (1982).

Alkaloids affect the central nervous systems of animals very dramatically, causing death in many instances. In fact, an extract of hemlock leaves was used by the ancient Greeks to murder the famous philosopher Socrates. Well-known alkaloids are opium, strychnine and nicotine. The latter is still used as an insecticide in many countries. On the other hand, certain insects have evolved mechanisms to cope with alkaloids and to detoxify them. Two Compositae weeds in England belonging to the genus *Senecio* contain a series of alkaloids in their tissues known as the pyrrolizidine type. These alkaloids are very effective in deterring herbivory and when accidentally ingested by cattle cause severe toxicity. In spite of this the caterpillars of the tiger moth and the cinnabar moth complete their entire life cycle on these plants. The moths sequester and store the alkaloids in their tissues and in turn use them as a chemical defence against bird predators. To enhance this protective mechanism they display suitable warning (aposematic) colours in both adults and larvae (Keeler, 1975).

Another fascinating example of the use of pyrrolizidine alkaloids is provided by male Danaidae butterflies, which transform these alkaloids into pyrroles that are stored in the wing hair pencils and are ultimately used as sexually stimulating pheromones. The male hovers over the female to stimulate her prior to mating. If the alkaloids are absent from the butterfly's diet the pheromone is not produced and courtship is not successful. For this reason the adult butterflies will suck plant sap from the withered leaves of *Senecio* plants to obtain the necessary precursor for the pheromone. This in itself is most unusual as most adult butterflies only require nectar or perhaps water during this period of their life cycle. But even more impressive is the ability of Danaidae butterflies to obtain these alkaloids from certain grasshoppers which release the plant sap they obtain from feeding on the leaves of *Heliotropium* plants. The chemical pathways illustrated in Fig. 3.2 show the different fates of pyrrolizidine alkaloids in mammals and insects and explain why they are toxic to the former and not to the latter (Edgar and Culvenor, 1974; Bernays *et al.*, 1977).

Cyanogenic glycosides liberate hydrogen cyanide (HCN) when fully hydrolysed and the latter is highly toxic to living tissue because it inhibits electron

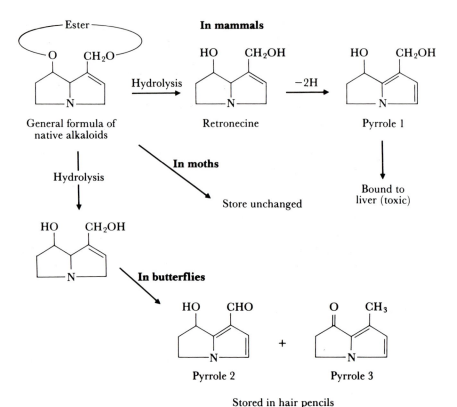

Figure 3.2   The metabolic fate of pyrrolizidine alkaloids in mammals and in insects. From Harborne (1982).

transfer in the cytochrome system, one of the most important life processes in living cells. Fairly high concentrations of cyanogenic glycosides occur in almonds, kernels of apricots, cherry pips and apple seeds. In fact, eating large amounts of apple seeds can be fatal in humans. Another more important hazard to humans is the cassava plant, the roots of which are used as a staple food in certain regions of West Africa. The roots are specially processed to reduce the HCN content before being turned into flour. Nevertheless, people relying on this plant as a staple food ingest a daily dose of about 35 mg HCN, which is approximately half the lethal dose. Not surprisingly, deaths do occur from eating cassava and life expectancies are not high on this diet. The latter phenomenon may, however, be also due to the goitrogenic effect of the non-toxic thiocyanates in the plant (Siegler, 1975). The *Asclepias* or milkweed family is well known for high concentrations of cardiac glycosides which protect this plant very effectively from herbivory. In one instance, however, a butterfly species *Danaus plexippus*, the well-known monarch, has actually exploited this defence mechanism for its own protection. The larvae feed on *Asclepias* species and are apparently not deterred by the bitter taste of the toxins. In fact it is their preferred food plant as they do not have many competitors. While feeding they sequester and store the cardiac glycoside in a harmless form and after metamorphosis the glycoside provides protection to the adult butterfly against its major predator, the blue jay. The adults also possess aposematic coloration which enhances the impact of the toxin upon the learning process of the blue jay. As a consequence non-toxic mimics have evolved but at least 50% of a similarly coloured butterfly population must possess the toxin for the aposematic coloration to remain effective (Harborne, 1982).

Acacia trees in Central America provide an unusual example of how secondary chemical defence can become redundant if other defence mechanisms are operating effectively. Many of these species have extra-floral nectaries that attract ant colonies, which then use the trees for shelter and food. Should any herbivore attempt to graze on the tree, they are attacked viciously by the ants. These *Acacia* trees contain no chemical defence but those without a symbiotic relationship with ants contain cyanogenic glycosides. To clinch matters, it has been shown that when ant colonies are artificially removed from young *Acacia* trees that normally harbour ants, these trees rarely survive beyond 6–9 months (Rehr *et al.*, 1973; Janzen, 1975).

Oils and resins are commonly used in trees to deter herbivory. They are also present in smaller plants, e.g. cannabis, the active principle of hashish, is a resin. Care should also be taken in selecting wood for barbecues, particularly in Africa, as some of the resins are volatilised at high temperatures and condense on the grilled meat. Severe vomiting and diarrhoea then ensue.

Oxalic acid is a very common deterrent and most Europeans will have tasted it in desserts made from the rhubarb plant. It has a very acid taste and after absorption it combines with ionised calcium (Ca^{2+}) in the blood plasma. The ensuing hypocalcaemia, particularly in the elderly, can be very debilitating if not fatal. In the arid areas of the United States the so-called locoweed (*Halogeton*) contains over 30% oxalic acid on a dry basis and has killed thousands of grazing sheep.

3.1.6 Functions of the nutrients

Volumes can and have been written on this subject. Our purpose here is merely to refresh a few elementary principles to allow the logical progression of subsequent ideas.

Carbohydrates are not essential nutrients in the diet of most animals. It is true that they make up the major portion of the diet of many animals and in doing so fulfil the very important function of providing energy. However, carnivores can exist without access to carbohydrates and rely either on lipids or deaminated amino acids for an energy source. Blood-sucking insects are a good example of high-protein feeders, while the traditional diet of some humans (e.g. the Inuit) was virtually carbohydrate free for most of the year. We must distinguish between highly digestible carbohydrates (NFE fraction of the proximate analysis) such as starches and sugars, and the crude fibre portion which consists mostly of cellulose and is difficult to digest. Both types of carbohydrate provide energy but special adaptations are required to digest cellulose, as we shall see.

Proteins, like carbohydrates, are not required as such by any animal. The essential amino acids contained in proteins are, however, required for a myriad of functions in animal cells. They are used to synthesise structural proteins in tissues as important as muscle, bone, hair, feathers, hoofs and horns as well as for the synthesis of delicate enzymes and hormones responsible for controlling the complex sequence of biochemical reactions within cells. They are called essential because the animal in question is unable to synthesise them and requires a dietary source of these amino acids (see section 3.1.3 for discussion on protein quality). A protein deficiency is far more likely to occur under natural conditions than any other deficiency, such as energy, and physiological ecologists should develop an acute awareness of the importance of this nutrient, particularly in deserts, prairies, steppe and savannah ecosystems.

The *essential fatty acids*, arachidonic acid, linolenic acid and linoleic acid, are required in the diet of all non-ruminant mammals studied thus far. They only occur in appreciable amounts in animal tissues. For this reason strict vegetarians (humans) run the risk of developing a deficiency in these essential nutrients. Deficiencies are slow to develop as there is a considerable storage in cell membranes and it is of interest to note that arachidonic acid is the precursor of the ubiquitous hormone prostaglandin.

The essential *macro-elements* such as calcium, phosphorus and magnesium function as structural elements as well as essential co-factors in metabolic pathways and the control of membrane function. Sodium, chlorine and potassium are essential electrolytes involved in osmoregulation and membrane function, whereas the essential *micro-elements* usually function as co-factors in enzyme systems. In the case of cobalt it is an integral part of the molecular structure of cobalamin or vitamin B_{12}.

The essential vitamins, both fat soluble and water soluble, are required in minute amounts and consequently, as one would expect, they function mostly in enzyme systems either as essential co-factors or as actual constituents of enzymes. The history of their discovery makes fascinating reading and is well worth the attention of readers preparing for a research career. For example, vitamin B_1, or more correctly thiamine, was discovered by a Dutch physician, Eijkman, working in the East Indies, who fed his chickens on polished and unpolished rice. The birds that received the polished rice soon developed fatal deficiency symptoms (beri-beri), whereas the other group flourished. Similarly, the English naval surgeon, Lind, discovered that sailors with access to limes (citrus fruit) during long voyages did not develop scurvy and this was the first step towards the discovery of vitamin C or ascorbic acid. The famous nineteenth-century explorer Captain James Cook forced his crew to eat sauerkraut regularly to prevent the development of scurvy. The dietary requirements for vitamins varies among animals. For example, insects require a dietary source of cholesterol, whereas mammals manufacture their own. Also guinea-pigs and primates appear to be the only animals that require vitamin C in their diet, and ruminant animals such as deer do not require a dietary source of the water-soluble B vitamins in their diet, because symbiotic micro-organisms in the rumen synthesise these vitamins for them.

3.1.7 Digestion

Because of the pre-eminent role that digestion plays in survival and the hedonistic pleasures of mankind, it is a physiological process which has attracted a great deal of both scientific and popular attention. Much is known about the delicate neural and endocrine control of the sequence of chemical events which occur during the passage of the ingesta through the digestive tract of mammals. The first hormone to be discovered was secretin and today a wealth of information on digestive physiology has accumulated. In this section, however, we shall assume a basic knowledge of the process and rather concentrate on certain comparative examples and their ecological significance.

Digestion in most animals is a carefully controlled process of molecular simplification whereby macromolecules of proteins, carbohydrates and lipids are reduced in a stepwise fashion to simpler molecules such as sugars, amino acids and fatty acids. This process obviously facilitates absorption of these nutrients but it also prevents powerful antigens like foreign proteins from reaching the general circulation. The latter function is not widely appreciated nor is it generally known that during heat stroke the integrity of the barrier, which the gut mucosa presents to the outside world, breaks down and toxins from bacterial digestion can invade the bloodstream causing fever, general septicaemia and death.

The detailed nature of the digestive process varies considerably among animals. For example, compare digestion in an internal parasite with that in a cow. In some animals such as crocodiles, some birds and even dinosaurs, stones are ingested and held in the so-called gizzard. When the strong muscular walls of the gizzard contract the stones move over one another producing the action of a stone mill on hard seeds and coarse plant material. These gastric mills are very effective and it has been shown that a tube of sheet iron, capable of withstanding 40 kg of pressure is completely flattened when introduced into the gizzard of a turkey. In Egypt, new stone carvings have been placed in bird gizzards to give them an antique appearance for

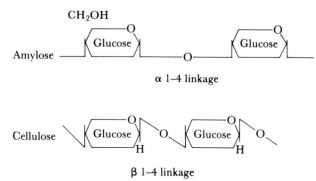

Figure 3.3 The different linkages between glucose molecules in starch and in cellulose.

later fraudulent sale to unsuspecting tourists visiting the Pharaohs' tombs. Digestion should, however, not be seen exclusively as a process of degradation. Many important compounds such as vitamins and amino acids are synthesised in the digestive tracts of a great variety of animals from termites to elephants. The special case of cellulose and ruminant digestion warrants further attention.

Cellulose, a constituent of the plant cell wall, is probably the most abundant and commonly available source of food energy available on our planet. This is true for all terrestrial biomes from Arctic lichen fields, through prairies to deserts, yet very few animals (snails, wood-boring beetles, shipworms, thysanurans) produce an endogenous cellulase enzyme, capable of reducing cellulose to its constituent glucose molecules. In contrast, starch, which is also constituted from glucose molecules, is highly digestible and the enzyme, amylase, which is responsible for starch digestion, is very widespread among animals. The difference between cellulose and starch would appear to be very minor and lies in the difference in chemical bonding between the glucose molecules (see Fig. 3.3). Nevertheless, this apparently small difference has led to the evolution of involved symbioses between micro-organisms and their animal hosts, in which the micro-organisms reduce the cellulose to volatile fatty acids for the hosts' benefit. It is quite remarkable that instead of evolving an enzyme to crack the β1–4 linkage in cellulose, a seemingly easy evolutionary step, animals such as antelopes, sheep, elephants, rhinoceroses, giraffes, horses, termites and many more have evolved complex digestive systems to accommodate the necessary micro-organisms which possess the enzymes to break the β1–4 linkage. In many instances the evolution of these attributes must have had a profound effect on the whole life style of these animals.

To illustrate some of the remarkable adaptations associated with cellulose digestion, let us consider ruminant digestion in some detail. Figure 3.4 shows the development of the digestive tract of a typical ruminant animal, the

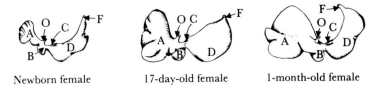

Newborn female   17-day-old female   1-month-old female

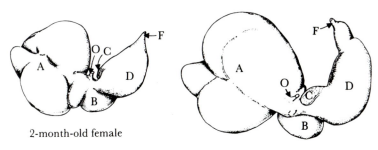

2-month-old female

3-month-old female

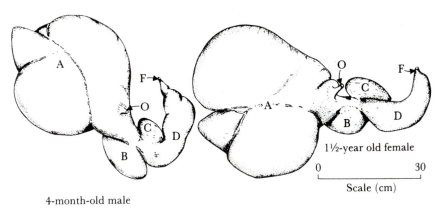

1½-year old female

0   30

Scale (cm)

4-month-old male

Figure 3.4   The development of the complex stomach of a typical ruminant, the white-tailed deer, A, rumen; B, reticulum; C, omasum; D, abomasum; O, oesophagus; F, duodenum. Redrawn from Moen (1973).

white-tailed deer. The newly born fawn is at first entirely dependent on milk for its nutrition and it would be entirely inappropriate, in fact harmful, if milk were to fill the undeveloped rumen because the rumen does not secrete any digestive enzymes. For this reason the milk is conveyed along a special groove, the oesophageal groove, directly to the true stomach or abomasum where it is digested in the normal way. Consequently the rumen, reticulum and omasum are bypassed. As the fawn develops it begins to eat increasing amounts of herbage and the rumen, reticulum and omasum become well developed and functional at the age of 3 months (Fig. 3.4). During this period of development the fawn is in close contact with the mother and cross-inoculation of the required rumen micro-organisms occurs.

Once the rumen is fully functional, the deer is able to consume large amounts of fibrous plant material. They feed rapidly and the food is hardly chewed before being swallowed. Once the rumen is full the animals retire to a safe area, usually with a favourable microclimate, and commence rumina-tion. The coarsest plant material in the rumen floats near the surface of the watery ingesta and a portion of it is returned to the mouth for further mastication by a process of regurgitation. This consists of closing the glottis and making a strong inspiratory effort. Try it yourself and you will feel your last meal moving in the direction of your oesophagus! After thorough chew-ing between the powerful and sharp cutting edges of the molars, the fine plant material is returned to the rumen where it tends to settle below the coarse, recently consumed material. The fine material is now exposed to microbial digestion.

The process of microbial digestion is largely one of fermentation but very important synthetic processes also occur. Conditions in the rumen are ideal for microbial growth: the temperature is constant at 37–40 °C; the pH is kept constant by a voluminous secretion of litres of saliva containing large quanti-ties of $NaHCO_3$; the ingesta is kept moist and continuous mixing of the contents by contraction of the strong muscular pillars in the rumen wall, all add up to an ideal fermenting apparatus. The major chemical changes which take place in the rumen are (1) the digestion of cellulose to the volatile fatty acids, acetic, propionic and butyric acids, (2) the synthesis of amino acids from non-protein organic nitrogen such as urea, and (3) the synthesis of the water-soluble B vitamins. These changes are naturally brought about by the microbes for their own benefit but eventually, as these millions of teeming organisms finally reach the true stomach or abomasum where digestive enzymes are secreted for the first time, they are killed, digested and utilised by the host animal. The volatile fatty acids are used as an important energy source and consequently blood levels of glucose are lower in ruminant mam-mals than in monogastric mammals.

The ability of ruminant animals to use coarse, apparently useless dry plant material and to transform it into such high-quality products as meat and milk, not to mention wool and the energy to act as efficient draught animals, is truly remarkable. It is therefore not surprising that the ruminants have undergone considerable adaptive radiation and are among man's most prized domestic animals. When next you view a springbok sprinting across the African plain or a Rocky Mountain goat jumping effortlessly between cliff faces, remember that the fuel for these remarkable movements is provided by a humble compost heat in the enlarged forestomach. To further emphasise the nutritional independence which ruminant digestion imparts upon an animal, consider any large antelope of the African savannah, e.g. the oryx.

For most of the year the African savannah is dry and the vegetation is high in cellulose, low in protein and vitamins, yet the oryx survives well under these conditions. The rumen microflora degrade the cellulose to fatty acids which are absorbed and used for energy. Instead of excreting all the urea in the urine the oryx recycles some urea, obtained from the deamination of amino acids in its tissues, into the rumen ingesta. The urea is used by the microbes to synthesise amino acids and in this way the minimum protein requirements of the oryx are largely met. The animal also obtains the vitamin B complex from the synthetic activity of the rumen micro-organisms, while vitamins K and E are synthesised by bacterial action in the lower digestive tract. Vitamin C is synthesised in the tissues, as is vitamin D from cholesterol under the influence of ultraviolet light. If the animal is ranging over a wide area, the parched pasture should still supply all the necessary macro- and micro-elements. In this regard cobalt can sometimes be deficient and this causes anaemia, ataxia and eventual death, as cobalt is essential for the rumen micro-organisms for synthesising cobalamin (vitamin B_{12}). Sodium also tends to be in short supply in nature and salt licks, either natural or artificial, will meet this requirement. What is left? Only one nutrient, namely vitamin A, and unfortunately this must be present in the diet as it cannot be synthesised. During the brief rainy season on the savannah, however, the grasses turn green and grow rapidly. At this stage of their growth they contain high concentrations of carotene (2 mol of vitamin A) and the oryx is then in a position to store large quantities of vitamin A in the liver which will last for many months. If we now return to the question posed at the beginning of this section as to why so many animals have evolved such elaborate mechanisms for digesting cellulose instead of merely synthesising the enzyme cellulase, we are in a better position to answer it. It seems, for example, that the synthetic processes which accompany cellulose digestion by symbiotic micro-organisms are just as important as the mere degradation of the cellulose and that these advantages have obvious survival value.

Ruminant digestion is naturally not the only method of digesting cellulose. Termites also use symbiotic micro-organisms in their digestive tracts for this purpose and are probably the most important herbivores on our planet. The amount of solar energy channelled through cellulose and then through digestion in termites is phenomenal. Ecologists have even gone so far as to suggest that the methane gas evolved from cellulose digestion in termites may one day be sufficient to alter the heat balance of our planet by influencing the familiar greenhouse effect. Polar ice caps could melt and coastal cities could be swamped! The possibilities for imaginative speculation bordering on science fiction are endless.

Many mammals that are able to digest cellulose have evolved caecal digestion instead of ruminant digestion. In this form of digestion the food is subjected to careful mastication prior to swallowing, as regurgitation does not occur. True gastric and duodenal digestion also occur before the ingesta reaches the enlarged caecum where its passage is slowed and microbial fermentation of cellulose occurs. This is quite different to ruminant digestion where fermentation precedes gastric and duodenal digestion. This difference is a disadvantage in the case of the caecal digestors, because the opportunity for absorption distal to the caecum is limited, whereas distal to the rumen there is ample opportunity for both digestion and absorption. This disadvantage is partially overcome in rodents by recycling their faeces, known as coprophagy. Rabbits produce special faecal pellets at night for consumption and it is thought that the major purpose in recycling faeces is to allow further processing and absorption of the fermentation products produced in the caecum. Coprophagy can, however, serve other purposes such as transferring pheromones in social rodent species. Important examples of caecal digestors are rodents and the equids (horses, zebras, donkeys, etc.).

Christine Janis (1976) has analysed the evolutionary implications of caecal and ruminant digestion in ungulates. She points out that caecal digestion is in fact a superior adaptation for dealing with really high levels of fibre in the diet as long as there is an unlimited quantity of herbage. She bases this view on the fact that horses will increase the rate of passage when challenged with very high concentrations of fibre in the diet, whereas ruminants are not able to do so. Under more favourable conditions, however, caecal digestion only reaches an efficiency of 70% of that of ruminant digestion. She also is of the opinion that perissodactyls developed caecal digestion of cellulose early in their evolution when they were still small. In contrast, the ruminant artiodactyls developed cellulose digestion only after they were of a sufficiently

(a)

(b)

(c)

Plate 3.1 Ruminant animals (a and b) are capable of digesting cellulose by means of symbiotic micro-organisms in their capacious forestomachs or rumens. Digestion of cellulose in equids (c) occurs in enlarged caecae after gastric digestion. Ruminant animals have undergone extensive adaptive radiation (speciation), whereas the equids have not. (Photo A: courtesy Barry Lovegrove.)

large size to accommodate this type of digestive tract. This attribute plus their ability to switch opportunistically to different plant taxa are the major reasons, according to Janis (1976), for the remarkable adaptive radiation of the ruminants when compared with the relatively low rate of speciation in equids. She might have added that the feeding pattern of ruminants, rapid prehension and swallowing followed by long periods of rest, allows these animals to seek shade or shelter and remain alert to predators, while the equids are obliged to feed almost continuously. This fundamental difference has influenced social behaviour and reproductive patterns and must be an important consideration in their evolution.

Digestion and absorption are, however, not confined to coping with large chunks of coarse and fibrous plant material. They can be very delicate processes indeed when one considers the absorption of dissolved organic material by marine invertebrates. Until recently there was considerable doubt about the net flux of primary amines into these animals because of significant losses of these amines which occurred concurrently. Wright (1987) reports, however, that this question has now been resolved and that marine mussels and echinoderms are able to absorb and accumulate dissolved free amino acids from external concentrations of less than $10 \, \mathrm{nmol \, l^{-1}}$. This concentration is well below the lower limit of concentration found in natural waters and represents a remarkably efficient form of absorption.

Although digestive processes appear to be reasonably similar in vertebrate animals one would expect that rates of absorption would differ in view of large differences in metabolic rate and general life style of these animals. This seems to be the case when we examine the data of Karasov (1987). Karasov measured the rate of uptake of glucose and proline in a large number of different species of fish, reptiles and mammals (see Fig. 3.5) and found that the uptake rate of these nutrients in the entire gut was 13 times higher in mammals than in fish and four times higher in mammals than in reptiles. He ascribes the faster rate of absorption in mammals to the greater surface area of the intestine in these animals, whereas the superior absorption in mammals and reptiles, when compared with fish, is ascribed to the higher operative temperature of the intestine in these two classes.

Diamond and Buddington (1987) have used the ratio of proline transport to glucose transport (P/G ratio) as an index of intestinal adaptations for absorbing protein or carbohydrate. Their extensive studies have included 32 vertebrate species from all of the five higher classes and show that the P/G ratio is highest for carnivores, intermediate for omnivores and lowest for herbivores. These differences are related to differences in the action of transport proteins. These authors also found that when prickleback fish undergo the developmental change from carnivory to herbivory, this transformation is not induced by a change in diet but appears to be endogenously triggered when the fish reach a specific age.

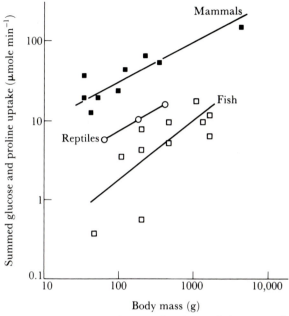

Figure 3.5 Relationship between the sum of proline and glucose uptake rates and body mass in fish, reptiles and mammals. Simplified from Karasov (1987).

3.2 Energy

The term 'energy' is a household word but few of us really understand the abstract complexities of this difficult concept. The student is again advised at this stage to review certain major physical principles associated with potential, kinetic, chemical, electrical, mechanical and radiant energy. The laws of thermodynamics are of universal importance and all animals must conform to them. The concept of entropy is also essential background knowledge for the ecophysiologist and is explained by the second law of thermodynamics, which states that matter is continuously and spontaneously decaying into a disordered state of randomness known as entropy. For example, when glucose is degraded to CO_2 and H_2O in an animal's tissues some of the energy is trapped as chemical energy in the high-energy phosphate bonds of ATP, but much is also lost as random kinetic energy in the form of heat, thereby increasing universal disorder. For this reason when physicists are asked to define life, they frequently describe it as a process or means to beat entropy, albeit only temporarily. In this section we shall examine the cost of 'beating entropy' in a variety of animals, the major factors influencing that cost and how to measure them. In other words we shall study the 'cost of living'.

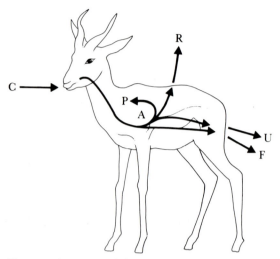

Figure 3.6 Theoretical concept of the major avenues of energy flow through a mammal. *C*, gross energy contained in food; *P*, energy deposited in and retained in tissues; *F*, energy lost in faeces; *U*, energy lost in urine; *R*, energy lost in respiration or as heat; *C − F = A* is the amount of assimilated energy.

3.2.1 Energy metabolism and the energy budget

The term 'energy metabolism', or 'metabolic rate', has developed into a catch-all term to describe the rate at which an animal uses energy to sustain essential life processes such as respiration, blood circulation, muscle tonus as well as for many other additional purposes, the most important of the latter being growth, reproduction, lactation and activity. An animal can therefore be viewed as a dynamic and open-ended flow-through system in which matter is transformed into work and heat. This concept can be stated more precisely in the form of the so-called energy budget (Figs 3.6 and 3.7).

Figure 3.6 shows that ingested food energy (C) is subjected to digestion but only a portion thereof (A) is actually assimilated, the undigested energy being lost as faecal energy (F). Some of the assimilated energy (A) may be deposited in the tissues and this represents growth or production (P). During the metabolic processes which accompany growth, however, a considerable amount of energy will be lost as heat (R) and smaller amounts lost when simple molecules are voided in the urine (U). This allows us to write a simple equation to describe the energy budget. Note that in many animals, e.g. insects and birds, $F + U$ are mixed in the terminal portion of the gut and are therefore expressed as one component.

$$C = P + R + U + F$$

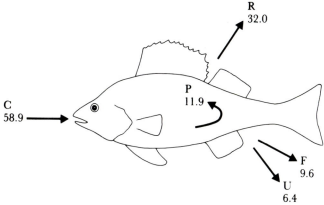

Figure 3.7 Energy partitioning (kJ) in a perch during a period of 28 days. Redrawn from Brafield and Llewellyn (1982).

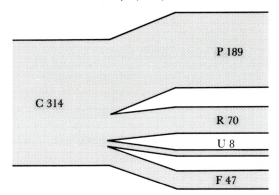

Figure 3.8 An estimate of energy flow (kJ m^{-2} year^{-1}) through a population of polychaete worms living in an inter-tidal mudflat. Redrawn from Kay and Brafield (1973).

An actual energy budget for a perch, based on the above equation appears in Fig. 3.7.

If the energy gains and losses depicted in Fig. 3.7 are added together and compared, it will be seen that the budget does not balance perfectly. This is due to inherent errors in the various techniques employed in these studies. It should also be noted that animals can exhibit negative energy balances when C drops below the minimum requirement for maintenance. In this case P will naturally become negative. When P is 0 the animal is in balance and the amount of energy in the food (C) is described as the maintenance ration.

Energy budgets are not restricted to individual animals and are frequently calculated for whole populations. For example, Kay and Brafield (1973) have estimated the flow of energy in a population of polychaete worms *Neanthes virens* living in an estuarine mudflat (Fig. 3.8).

The ultimate aim of many ecologists is, however, to use individual and population energy budgets to construct energy budgets for whole ecosystems. These efforts have led to the development of the concept of ecological efficiency and Brafield and Llewellyn (1982) have proposed the following measures of ecological efficiency: the energy coefficient of the first order (P/C); the energy coefficient of the second order (P/A); and the assimilation efficiency (A/C). In other words P/C is an estimate of how much energy, consumed by animals within a specific trophic level, is available for transfer to the next trophic level. P/A indicates how efficiently extracted food energy is converted into biomass and A/C is merely the efficiency with which energy is digested and absorbed from the food.

The most important factor influencing A/C will naturally be the nature of the food. Herbivores that have to cope with large amounts of cellulose exhibit relatively low A/C values ($c.\,57\%$), whereas carnivores that habitually feed on highly digestible food have A/C values of approximately 80%. Also, as we have seen in section 3.1.8, homeotherms are more efficient digesters than poikilotherms because of the higher constant temperature at which the gut is kept. In this respect it has been shown that growing varanid lizards, kept at sup-optimum temperatures and prevented from thermoregulating behaviourally, utilised their energy far less efficiently than those allowed to thermoregulate normally (Buffenstein and Louw, 1982). After the energy has been assimilated the efficiency of P/C is solely dependent on how much energy is actually laid down as tissue or biomass (i.e. P/A). Brafield and Llewellyn (1982) have assembled ecological efficiency data, based on the above concepts and drawn from a large variety of animals and sources. These data are contained in Tables 3.1 and 3.2.

The measurement of metabolic rate and energy budgets for individuals and populations is one of the most important aspects of ecophysiology because it is probably the most sensitive and precise assessment one can make about the function or impact that individuals and populations can have within a specific ecosystem. These measurements are in fact estimates of the total energy exchange of animals with the environment. For this reason energetic studies are expanding into many disciplines and are even being used to make educated guesses about our recent and ancient history. How, for example, was sufficient food energy made available for the thronging crowds of slaves, used to build the pyramids? Who were responsible for carving the massive stone heads on Easter Island, and where was the energy obtained for the labourers on this small and isolated island? The use of energetic studies cannot naturally answer all these questions, but they do help us to refine our guesses.

3.2.2 Oxygen, the fuel for energy release

The majority of animals rely on oxygen to release the full potential energy

Table 3.1. Individual animal energy efficiencies (as percentages) (after Schroeder, 1980). A, assimilation; P, production; C, consumption. Figures in parentheses show numbers of species involved. From Brafield and Llewellyn (1982)

			A/C	P/A	P/C
Poikilotherms (heterotherms)					
Herbivores	Terrestrial	(32)	45	52	20
	Aquatic	(15)	61	56	34
Granivores	Terrestrial	(4)	78	30	24
Carnivores	Terrestrial	(11)	84	58	46
	Aquatic	(17)	64	48	30
Detrivores	Terrestrial	(6)	12	50	6
	Aquatic	(6)	44	56	25
Parasites		(3)	77	50	42
Homeotherms					
Herbivores	Terrestrial	(3)	66	23	13
Granivores	Terrestrial	(3)	76	29	22
Lactivores	Terrestrial	(2)	95	45	43
Carnivores	Terrestrial	(–)	—	—	—

Table 3.2. Population energy efficiencies (as percentages) over variable time spans. Symbols as in Table 3.1. Figures in parentheses show number of species involved. From Brafield and Llewellyn (1982)

		A/C	P/A	P/C
Poikilotherms (heterotherms)				
Herbivores	Terrestrial	32 (5)	37 (69)	14 (4)
	Aquatic	47 (4)	26 (4)	11 (5)
Granivores	Terrestrial	1 (1)	—	—
Carnivores	Terrestrial	77 (2)	36 (19)	30 (2)
	Aquatic	—	—	—
Detrivores	Terrestrial	35 (3)	13 (1)	6 (1)
	Aquatic	45 (1)	32 (3)	6 (1)
Parasites		—	—	—
Homeotherms				
Herbivores	Terrestrial	65 (10)	3 (16)	2 (8)
Granivores	Terrestrial	81 (1)	4 (2)	4 (1)
Lactivores	Terrestrial	—	—	—
Carnivores	Terrestrial	88 (4)	2 (8)	2 (4)

contained in nutrient molecules such as fats, proteins and carbohydrates. This process involves a delicately controlled sequence of biochemical reactions which culminate in the stripping of hydrogen atoms from pyruvic acid during the Krebs (or tricarboxylic acid) cycle, the combination of oxygen with hydrogen in the cytochrome system and the synthesis of energy-rich adenosine triphosphate (ATP). Nevertheless, many animal species are entirely or partially dependent on anaerobic respiration. For example, the huge pogonophron worms which are thicker than a man's arm and live at great depths on the sea floor close to hydrothermal vents, change sulphides to elemental sulphur when oxygen is limited by means of endosymbiotic bacteria. Later in this chapter we shall also see that certain vertebrates, including amphibians and reptiles, rely quite heavily on anaerobic metabolism, particularly during periods of intense activity. Human athletes also employ anaerobic metabolism during short sprint events and in these cases nutrients such as glucose are metabolised to lactic acid, which accumulates in the tissues until the oxygen debt is repaid. During repayment of the oxygen debt the lactic acid is changed to pyruvic acid which then enters the Krebs cycle in the usual way. An extensive review of anaerobic metabolism is contained in Hochachka's (1980) text, *Living Without Oxygen*.

Both oxygen and carbon dioxide diffuse passively across the outer surfaces of organisms and, consequently, the rate of diffusion will be governed by similar physical factors that control the transfer of heat, solvents and solutes; namely the concentration gradient, the diffusion coefficient, the diffusion distance, as well as the surface area of the membrane involved in the diffusion process. These factors can be summarised in the following equation:

$$D_t = \frac{D_c A (C_1 - C_2)}{D_d}$$

where D_t is the rate of diffusion, D_c the diffusion coefficient, D_d the diffusion distance, A the surface area of membrane, and $(C_1 - C_2)$ the concentration gradient across the membrane.

It is also important to note that the concentration of an individual gas within a gas mixture will determine the *partial pressure* (P) of that gas. More precisely stated in terms of Dalton's law, this means that the partial pressure of an individual gas within a gas mixture is independent of the remaining gases present in the mixture, and the total pressure of the mixture is equal to the sum of the partial pressures of all the gases present in the mixture. For example, if oxygen occurs at its normal concentration of 20.94% in the earth's atmosphere and the atmospheric pressure is equal to 760 mmHg, then the partial pressure of O_2 is 159 mmHg. Figure 3.9 provides an informative overview of the availability of oxygen in typical aquatic and terrestrial habitats in terms of the partial pressure of oxygen. These data show the great variety of oxygen partial pressures present in our environment and highlight the need for special respiratory adaptations at the extremely low partial pressures experienced in mudflats and at very high altitudes. For example, the PO_2 at sea level is approximately 160 Torr and falls to 35 Torr at an altitude of 10 km. Figure 3.9 also shows that the percentage of O_2 in aquatic systems is far lower (0.5–1%) compared to the content in air (20.94%). This means that the first terrestrial animals immediately enjoyed a great metabolic advantage over their aquatic ancestors. Not only was the O_2 content of air

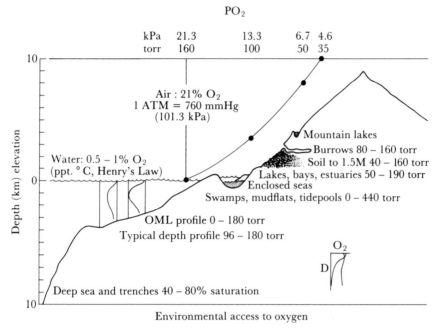

PO_2

| kPa | 21.3 | 13.3 | 6.7 | 4.6 |
| torr | 160 | 100 | 50 | 35 |

Air : 21% O_2
1 ATM = 760 mmHg
(101.3 kPa)

Mountain lakes
Burrows 80 – 160 torr
Soil to 1.5M 40 – 160 torr
Lakes, bays, estuaries 50 – 190 torr
Enclosed seas
Swamps, mudflats, tidepools 0 – 440 torr

Water: 0.5 – 1% O_2
(ppt. ° C, Henry's Law)

OML profile 0 – 180 torr
Typical depth profile 96 – 180 torr

O_2

D

Deep sea and trenches 40 – 80% saturation

Depth (km) elevation

Environmental access to oxygen

Figure 3.9 Availability of oxygen in various habitats. OML, oxygen minimum layer. Redrawn from Graham (1990).

far higher than water but because the density of air is so much lower than water, it became far easier to ventilate the respiratory surfaces in air than in water. These advantages eventually led to the evolution of endothermic birds and mammals with field metabolic rates as much as 13 times greater than those of ectotherms.

A great deal of research has been carried out on the respiratory adaptations exhibited by animals in a great variety of habitats. These include the evolution of specially enlarged respiratory surfaces, specialised respiratory enzymes and pigments, as well as the adaptations of ventilation patterns. In addition, the emergence of animals on land required that they also develop adaptations to minimise water loss and to excrete nitrogen in the form of urea and/or uric acid. The comparative physiology of respiration has been well covered in standard physiology texts such as Schmidt-Nielsen (1983), Eckert and Randall (1983) and in specialised reviews, for example Weibel *et al.* (1987) and Pierre Dejours (1989). We only have space to touch on three illustrative examples.

Comparative physiologists have long been fascinated by the ability of air-breathing mammals such as whales, porpoises and seals to undertake deep dives for remarkably long periods. The Antarctic Weddell seal is able to dive to a depth of 600 m and remains under water for more than 1 hour (Kooyman, 1981). Seals do not possess a larger lung volume than similarly sized non-diving mammals. In fact they collapse their lungs just prior to diving to minimise the air volume that can enter the bloodstream and tissues at the high pressures experienced at great depths. In this way they reduce the possibility of the bends developing when they return to the surface. Seals do, however, possess an increased blood volume and a slightly higher O_2 carrying capacity of the blood. Seals are able to store far more O_2 in their muscles because of an increased muscle mass, which also has a significantly higher myoglobin content. They are also able to tolerate a large O_2 debt. In addition, various circulatory changes take place in seals during dives, such as the slowing of the heart rate (bradycardia) while blood is shunted away from the periphery towards the central nervous system and heart. This pattern of adaptations illustrates an important concept, namely that adaptation to a highly stressful situation is often accomplished by a series of comparatively small adaptations acting in concert rather than a single very dramatic adaptation. The same principle could also be applied to the adaptation of animals to an extreme desert environment. For a more complete account of the physiology of diving animals see Schmidt-Nielsen (1983).

From ocean depths we proceed to geese flying over the Himalayas. How are these birds able to provide sufficient oxygen to their flying muscles at these great altitudes? The answer lies partially in the efficient avian respiratory system but this is not a sufficient explanation because high-altitude respiration is not a general characteristic of birds. Braunitzer and Hiebl (1988) have examined the molecular structure of the haemoglobin of the wild goose *Anser indicus*; individuals of this species have collided with aircraft at altitudes as high as 9500 m over India. Their results show that adaptation to high altitudes, or hypoxic stress, in this species is largely the result of a specific mutation in haemoglobin configuration; sequencing of this haemoglobin has revealed that the exchange of only one amino acid is involved in the mutation. This modified haemoglobin is then theoretically able to transport sufficient O_2 to the flight muscles of *Anser indicus* even at altitudes up to 12 200 m.

Although the preceding two examples illustrate significant adaptations among air-breathing vertebrates, Graham (1990) is of the opinion that aquatic animals display a much greater variety of respiratory adaptations. These range from the loss of haemoglobin in ice fishes because of the high solubility of O_2 in very cold waters, to the independent evolution of haemoglobin synthesis in many aquatic invertebrates exposed to hypoxic conditions. Graham (1990) concludes that several features of aquatic respiration distinguish it from aerial respiration, namely the frequent occurrence of integumental respiration, the frequent combination of respiratory and feeding surfaces, and the very significant effect of hypoxia in shallow water and the deep sea upon the development of specialised respiratory adaptations.

3.2.3 Methods for measurement of metabolic rate

We have already defined the essentially medical term of basal metabolic rate and shown that it is seldom possible to measure metabolic rate in animals when they are within their thermoneutral zone, in a post-absorptive state and at complete rest. Instead we describe the conditions under which metabolic rate was measured and define it as standard metabolic rate (SMR) or if the animal was at rest as resting metabolic rate (RMR).

The most obvious and one of the most accurate ways to measure RMR is by measuring the total heat produced by the animal in some form of calorimeter. This is known as *direct calorimetry* and was in fact the very first method used by the famous French physiologist Lavoisier some 200 years ago. Lavoisier's calorimeter was an ice calorimeter and consisted of a small chamber enclosed with double or cavity walls. The cavity walls were filled with ice and, after the subject was placed in the chamber, the body heat of the subject caused some of the ice to melt and the resulting water was weighed. This allowed the calculation of the total amount of heat produced by the subject per unit time by allowing for the latent heat required to produce a specific mass of water from ice. This type of calorimeter, although responsible for extremely important discoveries, was not suitable for refined experimentation for various reasons, including the cold to which the subject was exposed. Consequently, various modifications were introduced including the use of water calorimeters by Pettenkofer and Voit in Bavaria (1866), which measured the temperature of water before it flowed through an elaborate system of pipes within the chamber and after it emerged from the chamber (Fig. 3.10). If the chamber was well insulated and the flow rate of the water was known, the difference in temperature between the incoming and excurrent water would allow calculation of the total heat production per unit time by the subject. This technique represented an important improvement but it remained clumsy with a very slow response time. Today, modern calorimeters consist of very well insulated chambers with numerous thermocouples strategically placed all over the inside walls. These thermocouples are interconnected in the form of a grid to form a thermopile and this in turn is connected to a computer and/or data logger to collate and monitor the data continuously. Calorimeters of this nature are very expensive to build and require expert technicians to maintain them. They do, however, have the important advantage of measuring total metabolism including anaerobic metabolism, because the latter process naturally also produces heat. Indirect calorimetry, on the other hand, is based on oxygen consumption or aerobic metabolism and excludes anaerobic metabolism.

Because of the high cost and low flexibility of direct calorimetry, most laboratories employ some form of indirect calorimetry and the most common technique in international use today is the measurement of oxygen consumption. The measurement of CO_2 production is less reliable because in some

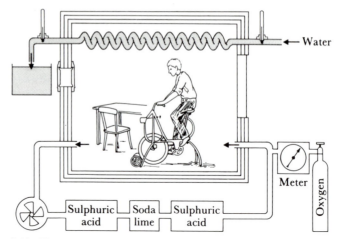

Figure 3.10 The Atwater human calorimeter. Redrawn from Durnin and Passmore (1967).

animals significant amounts of CO_2 are stored in the blood and tissues. Ideally both O_2 consumption and CO_2 production should be measured simultaneously. This allows calculation of the RQ and therefore identification of the substrates being metabolised. This, in turn, allows the researcher to assign a more accurate energy value to the amount of oxygen consumed. Indirect calorimetry is based on the assumption that the catabolic products of proteins, carbohydrates and fats ultimately join the common tricarboxylic acid cycle and consume O_2 while producing CO_2, water and heat:

$$C_6H_{12}O_6 + 6O_2 \rightarrow 6H_2O + 6CO_2 + \text{energy}$$

Using molecular mass and Avogadro's law that 1 g molecule of gas occupies 22.4 l then:

$$180\,g + 6 \times 22.4\,l = 6 \times 18\,g + 6 \times 22.4\,l + 633\,kcal$$

If we now divide by 6×22.4 then 1 litre of O_2 will combust 1.34 g glucose producing 1 l of CO_2 and 4.93 kcal. The RQ in this case is equal to 1 but in the case of fats with many more hydrogen atoms per unit mass the ratio approaches 0.7. It is clear therefore that the RQ is essential for great accuracy in assigning energy values to volumes of oxygen consumed. Nevertheless many, if not most, animals are usually catabolising a mixture of substrates and Durnin and Passmore (1967) maintain that the use of the following compromise formula allows the researcher to measure only O_2 consumption with an error no greater than $\pm 0.5\%$:

$$\text{Energy value (kcal min}^{-1}) = \frac{4.92 \times \dot{V}}{100}(20.93 - O_{2e})$$

Table 3.3. The relationship between heat produced, oxygen consumed and the respiratory quotient (RQ) when the major nutrients are catabolised. From Schmidt-Nielsen (1983)

Food	kcal g^{-1}	$1 O_2$ g^{-1}	kcal l^{-1} O_2	$RQ = \dfrac{CO_2 \text{ formed}}{O_2 \text{ used}}$
Carbohydrate	4.2	0.84	5.0	1.00
Fat	9.4	2.0	4.7	0.71
Protein (urea)	4.3	0.96	4.5	0.81
Protein (uric acid)	4.25	0.97	4.4	0.74

where $\dot{V}$ is the volume of expired air in litres per minute at standard temperature and pressure, 20.93 is the percentage of oxygen in normal air and O_{2e} is the percentage of oxygen in the expired air. For the sake of completeness the energy values for all substrates is given in Table 3.3. We can now turn our attention to the actual measurement of oxygen consumption.

* Note that the most commonly used units are:
    (1) one calorie = amount of heat required to raise the temperature of 1 g of pure water by 1 °C at 15 °C
    (2) one newton = force required to accelerate mass of 1 kg 1 m s^{-1} s^{-1}
    (3) one joule = work performed when point of application of 1 newton is displaced 1 metre in direction of force
    (4) one joule = 0.239 calories
    (5) one watt = 1 joule s^{-1}

The simplest way in which to measure oxygen consumption is to place an animal in an enclosed container with a suitable CO_2 absorbent and attach the container to a manometer (U-tube filled with fluid). As the animal uses oxygen, the volume of air in the container is reduced and this is registered on the manometer. The manometer can be calibrated by injecting air into the closed container with a graduated syringe. These so-called manometric techniques can, if care is taken, give accurate and reliable results. For example, an empty chamber (thermobarometer), an exact replica of the experimental chamber, should also be used to measure the effects of changes in temperature and pressure during the course of the experiment. These values are then either added to or subtracted from the experimental data, depending on which direction the changes occur. All volumes must first be transformed to conditions of standard temperature and pressure by using Charles' law before being reported:

$$\text{Volume (at STP)} = V_{\text{exp}} \times \frac{BP_{\text{exp}} \times 273}{760 \times T_{\text{exp}} + 273}$$

where BP is barometric pressure, V the gas volume and T the temperature.
    The above manometric techniques are naturally of limited value as they

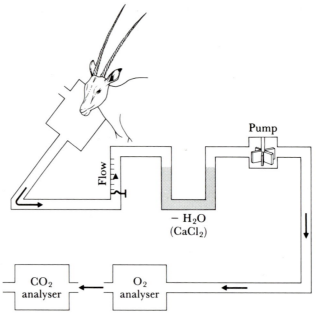

Figure 3.11   An open flow-through system for measuring oxygen consumption. See
text for details.

can only be used for small animals. Moreover, the animal cannot indulge in intense activity and the result is always total oxygen consumption over a fairly long period. The latter disadvantage is a very real one, as the investigator frequently wished to associate a specific kind of activity with the O_2 consumption at that moment. For this purpose an almost instantaneous reading is required and this can be obtained by using an open flow-through system as depicted in Fig. 3.11. The animal is fitted with a light mask and normal ambient air of a predetermined oxygen concentration is pumped through the mask, flow meter and drying agents, before being sampled and analysed by either an O_2 analyser or CO_2 analyser or both. Because the flow rate of air is known as well as the difference in O_2 and CO_2 content between the incoming air and the excurrent air, the O_2 consumption and CO_2 production by the animal can be calculated. This technique allows the animal to indulge in a reasonable variety of activities, including running on a treadmill with the added and important advantage of an almost instantaneous measurement of O_2 consumption.

Similar systems have been designed for aquatic animals using suitable flow metres and oxygen sensors for measuring the O_2 content of water. All these systems, however, suffer from the major disadvantage that they cannot be used to study free-ranging animals under natural conditions. Consequently, many techniques have been evaluated to attain this elusive goal. These have included the radiotelemetric measurement of heart rate and respiration rate, as well as the rate of disappearance of various isotopes from the body, i.e. ^{131}I, ^{85}Sr, ^{47}Ca, ^{65}Zn, ^{32}P. None of these techniques has given satisfactory results, apart from the use of ^{22}Na to measure food intake, and they have now all been supplanted because of the excellent success obtained with the doubly labelled water (DLW) method. This technique has been validated in a large variety of animals and, apart from insects, is the method of choice for measuring the energetic cost of free existence in free-ranging, terrestrial animals. Let us consider first the principles upon which it is based and then its actual use.

The use of doubly labelled water ($D_2{}^{18}O$ or $^3H^{18}OH$) is based on the experimental evidence that when $H_2{}^{18}O$ is injected into either rats or mice, the oxygen in exhaled CO_2 is in isotopic equilibrium with the oxygen of body water. Therefore a decline in the ^{18}O content of body water in an injected animal will be related over time to the rates of CO_2 loss as well as body water loss. Also, if the body water is labelled with a hydrogen isotope, the kinetics of the hydrogen isotope is related only to water loss, whereas oxygen is lost both through H_2O and CO_2. The difference in turnover rates of the two isotopes is therefore a function of the rate of CO_2 production.

The $D_2{}^{18}O$ method is therefore an estimate of CO_2 production and its use is based on the following assumptions:

1. Body water volume must remain constant.
2. The isotopes must label only the CO_2 and H_2O in the body.
3. Neither labelled nor unlabelled CO_2 or H_2O in the environment must enter the skin or respiratory surfaces of the experimental animals.

It is reasonably easy to comply with the conditions required by the above assumptions, except under very hot conditions when evaporative water loss can be excessively high and reduce the volume of body water. When dealing with fossorial animals, particularly social mammals, the third assumption cannot be met. The method can also only be used on animals that can be easily captured, injected with $D_2{}^{18}O$, liberated for several days/weeks and recaptured again for sampling of the body fluids. After the first injection a period of time (several hours) is allowed for the isotopes to equilibrate before the first blood sample is taken, prior to release. Ideally this equilibration time should be predetermined in a separate experiment. It is not surprising then that nesting birds that return faithfully to their nest sites are good subjects for DLW studies. The only major drawback to the method is the high cost of ^{18}O. This can be partially overcome by using charged particle activation of ^{18}O in a cyclotron, before measuring activity in any of the samples. The activation process transforms ^{18}O to ^{18}F, which is a gamma emitter and can be more easily measured at low concentrations than required by older techniques. Modern isotope ratio mass spectrometers are, however, so sensitive that only very small doses of ^{18}O are required. In this way costs can be reduced by initially injecting low concentrations of ^{18}O into the animals. Tritium is inexpensive and easily measured in a scintillation counter, while deuterium is a stable isotope and requires mass spectrometry for its measurement. The DLW method with many of its results has been thoroughly and critically reviewed by the leading international expert on this subject, Ken Nagy of the University of California in Los Angeles (Nagy, 1980, 1983, 1987; Nagy and Peterson, 1987). Before leaving this topic of DLW it should also be noted that Bradshaw *et al.* (1987) of the University of Western Australia in Perth have developed an alternative method of measuring DLW in body fluids involving prompt nuclear reaction analysis. It shows great promise and close agreement with Nagy's methods.

3.2.4 Scaling metabolic rate

The relationship between metabolic rate and body mass has provided both physiologists and ecologists with endless opportunities for speculative discussion but many fundamental questions have still to be resolved. A most readable and informative review of scaling in general has been written by Schmidt-Nielsen (1984). A similar review, but more ecologically orientated, has been published by Peters (1983). Both of these reviews are strongly recommended to the interested student.

The first attempt to examine the scaling of metabolic rate seriously was by Max Kleiber who published his now classical mouse to elephant curve in 1932. Based on this curve, Kleiber formulated an allometric equation to provide the best fit for his data:

$$P_{met} = 73.3\, M_b^{0.74}$$

where P_{met} is the metabolic rate in kcal day^{-1} and M_b the body mass in kg.

The most important first conclusion to be drawn from Kleiber's finding is that the slope of the regression line is equal to 0.74 and not 0.67 (Fig. 3.12). The latter would be expected if metabolic rate scaled to the same exponent as the surface area of the body. This is important because, until Kleiber published his findings, the non-linear relationship between metabolic rate and body mass was thought to be due to non-linear differences in heat loss, which in turn are a function of relative surface area. For example, the famous nineteenth-century German physiologist, Rubner, while working on different sized dogs, concluded that smaller dogs would lose more heat as a result of their greater relative surface area and consequently they must produce more heat to compensate for this loss. He argued that because the relationship between body size and surface area is non-linear, this provided the explanation for the relationship between body mass and metabolic rate.

Rubner's arguments were very convincing at the time because the difference between a slope of 0.67 and 0.75 was sufficiently small to be attributed to experimental error. In fact, far more recently Heusner (1982) proposed

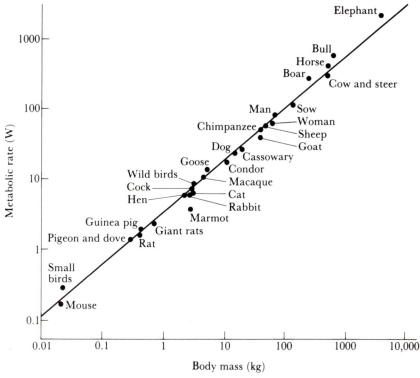

Figure 3.12 Straight-line relationship between log metabolic rate and log body mass in mammals and birds. After Benedict (1938) and Schmidt-Nielsen (1984).

that the exponent of 0.75 was a statistical artefact obtained from lumping species together from too great a range in body masses. He argued further that when exponents were calculated within species they were closer to 0.67 than to 0.75 (Fig. 3.13). Heusner's proposal has, however, been challenged by Feldman and McMahon (1983). Schmidt-Nielsen (1984), in reviewing the mathematical analyses of the latter authors, concludes that the exponent 0.75 is still the most valid statistical description for interspecific comparisons among mammals. When this relationship is plotted as a mass-specific or relative metabolic rate against body mass on a double log plot, the regression line has a slope of −0.25.

It seems therefore that we must reject Rubner's explanation at this stage, not only because the accepted exponent for describing the relationship between metabolic rate and body mass is not 0.67, but also because many ectothermic animals exhibit a very similar exponent to mammals. In the case of ectotherms they do not have to produce additional heat to compensate for increasing heat loss as they become smaller, with relatively larger surface

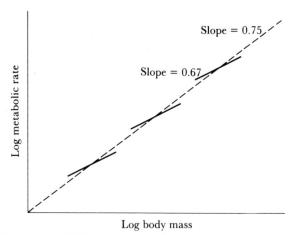

Figure 3.13   Heusner (1982) maintains that the regression line (– – –) with a slope equal to 0.75 is a statistical artefact, as a result of including too many individual species which in fact exhibit individual slopes of 0.67 (——). Redrawn from Schmidt-Nielsen (1984).

areas. The latter is the most persuasive argument against the surface area hypothesis. Nevertheless, surface areas probably do play some role in this interesting relationship. One need only remind readers that if a moose had the same metabolic rate as a mouse, the temperature of a moose's skin surface would be close to the boiling point of water. Moreover, the surface area of the body is not the only variable which changes with body mass, the number of cells changes and their total and relative surface areas in relation to total body mass also change. It needs little imagination to realise how the many energy exchanges on the membranes of cells could influence the fairly crude measurement of total metabolic rate.

Many more criticisms can be advanced about the blind acceptance of a slope of 0.75 as being the final answer to the correct slope for scaling metabolic rate. Very variable results have, for example, been obtained with invertebrates and this is not unexpected when we compare the rich variety of life styles and mechanisms of locomotion among these animals. Also, as Schmidt-Nielsen (1984) points out, merely by choosing one set of data, one arrives at a slope of 1.0 for colubrid and boid snakes, whereas another data set provides a slope of 0.6 for the same snakes. We may conclude therefore that many exceptions exist to the use of a slope of 0.75 to scale metabolic rate. Nevertheless, it is a very acceptable generalisation and, if actual measurements of a particular species are not available, it is a most useful tool for partially refining some ecological studies. Also, from a theoretical point of view, explanations of the various concepts involved in scaling metabolic rate still provide many unanswered questions and research opportunities. For

Plate 3.2   The mass-specific metabolic rate of the smallest flightless bird in the
              world, the flightless rail (40 g) from Inaccessible Island, is nine times that
              of the largest flightless bird in the world (90 kg), the ostrich. See text for
              discussion on scaling of metabolic rate. (Photo: courtesy Peter Ryan.)

comprehensive reviews see Hemmingsen (1960) and Schmidt-Nielsen
(1984).

The previous discussion on scaling has been limited to SMR, which is

Table 3.4. Allometric equations for calculating FMRs and feeding rates of free-living mammals, birds, and lizards. The equations have the form $y = ax^b$ where y is FMR (in kJ day^{-1}) or feeding (dry matter ingestion) rate in (g day^{-1}) and x is body mass (in g). From Nagy (1987)

Group	Units of y	a	x	b	95% CI of predicted y, as % of predicted y	Equation
Eutherian mammals						
All eutherians	kJ day^{-1}	3.35	g	0.813	-58 to $+138\%$	18
	g day^{-1}	0.235	g	0.822	-63 to $+169\%$	19
Rodents	kJ day^{-1}	10.5	g	0.507	-52 to $+110\%$	20
	g day^{-1}	0.621	g	0.564	-64 to $+176\%$	21
Herbivores	kJ day^{-1}	5.95	g	0.727	-62 to $+161\%$	22
	g day^{-1}	0.577	g	0.727	-62 to $+161\%$	23
Desert eutherians	kJ day^{-1}	3.21	g	0.786	-59 to $+141\%$	24
	g day^{-1}	0.150	g	0.874	-52 to $+108\%$	25
Marsupial mammals						
All marsupials	kJ day^{-1}	11.8	g	0.576	-42 to $+72\%$	26
	g day^{-1}	0.492	g	0.673	-37 to $+59\%$	27
Herbivores	kJ day^{-1}	6.36	g	0.644	-40 to $+67\%$	28
	g day^{-1}	0.321	g	0.676	-46 to $+84\%$	29
Birds						
All birds	kJ day^{-1}	10.9	g	0.640	-57 to $+135\%$	30
	g day^{-1}	0.648	g	0.651	-55 to $+124\%$	31
Passerines	kJ day^{-1}	8.88	g	0.749	-53 to $+111\%$	32
	g day^{-1}	0.398	g	0.850	-31 to $+45\%$	33
Desert birds	kJ day^{-1}	5.05	g	0.660	-47 to $+91\%$	34
	g day^{-1}	1.11	g	0.445	-49 to $+95\%$	35
Seabirds	kJ day^{-1}	8.01	g	0.704	-53 to $+113\%$	36
	g day^{-1}	0.495	g	0.704	-61 to $+159\%$	37
Iguanid lizards						
All iguanids	kJ day^{-1}	0.224	g	0.799	-32 to $+46\%$	38
Herbivores	g day^{-1}	0.019	g	0.841	-59 to $+146\%$	39
Insectivores	g day^{-1}	0.013	g	0.773	-30 to $+43\%$	40

measured under artificial conditions in the laboratory on resting animals. Of perhaps greater relevance and significance for the ecophysiologist is to study the scaling of field metabolic rate (FMR). This is measured by means of doubly labelled water under natural conditions in free-ranging animals (see section 3.2.2). Nagy (1987) has assembled data on the FMR of 23 species of eutherian mammals, 13 species of marsupial mammals and 25 species of birds. He found that FMR was closely correlated with body mass in each of these groups but that FMR scales differently to SMR in eutherians (0.81) and marsupials (0.58) but is similar to SMR in passerine birds (0.75). Of great interest to the physiological ecologist is Nagy's conclusion that medium-sized (240–550 g) eutherians, marsupials and birds have similar FMRs, but these are about 17 times higher than FMRs of similarly sized ectothermic vertebrates such as iguanid lizards. This vast difference in the energetic cost of free existence is of great ecological and evolutionary importance. Using the same data, Nagy (1987) has also provided us with a series of most useful equations to predict the FMRs and feeding rates of a variety of different animal groups (see Table 3.4).

Nagy (1987) explains the use of these predictive equations as follows: assume that we wish to estimate the annual cost of free existence in a spotted skunk (*Spilogale putorius*) and that the males weigh 900 g and the females 500 g and that there are 3 pairs ha^{-1} of habitat. From equation 18 in Table 3.4 we can predict an FMR of 845 kJ day^{-1} for the males and 524 kJ day^{-1} for the females. Annual energetic costs would be 308 MJ year^{-1} for the male and 191 MJ year^{-1} for the female. Energy respired by the population would be 1497 MJ ha^{-1} year^{-1} [(3 × 308 MJ year^{-1}) + (3 × 191 MJ year^{-1})]. Equation 19 now allows us to predict the dry matter (food) requirements at 63 g day^{-1} for the male and 39 g day^{-1} for the female. Annual food requirements for maintenance would be 23 kg year^{-1} for the male, 14 kg year^{-1} for the female and 111 kg ha^{-1} year^{-1} for the population. A correction factor of 1–3% should, according to Nagy, be applied to compensate for biomass in weaned offspring and growth of tissue in the adults. It should be noted that if the metabolisable energy value of the diet is known from feeding experiments, then feeding rates can be calculated directly by dividing FMRs by the metabolisable energy of the food in kJ g^{-1}.

3.2.5 Factors affecting energy requirements

Activity
When an animal is momentarily at rest in a thermoneutral environment, the energy it requires is referred to as the maintenance energy, as it is just sufficient to maintain body mass and the vital functions. The difference or range between the RMR and the maximum rate of energy expenditure, which the animal is capable of when engaged in maximum activity, is known as the metabolic scope. Bartholomew (1982) refers to this range as the factorial scope and provides the following interesting values for the ratio of minimum to maximum metabolic rates: chipmunk 7×, lemming 3×, birds 10× and moths 150×. These figures do, however, exclude anaerobiosis.

The huge differences between insects and birds raises various interesting questions including the theoretical maximum and minimum sizes of these two groups of animals. Physical principles inform us that the metabolic power for flight must increase with body mass but, because of anatomical constraints, the metabolic energy available to power flight increases at a lower rate. Therefore a threshold value for maximum body mass must exist beyond which flapping flight will not provide sufficient lift. This seems to be the case and the heaviest bird capable of flapping flight is the Kori bustard of

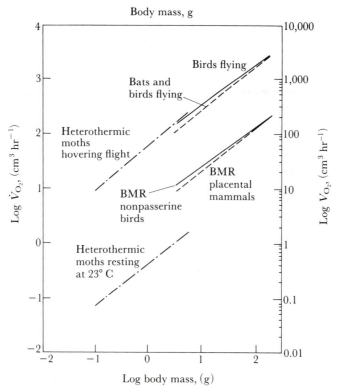

Figure 3.14   Relationship between log oxygen consumption and log body mass in insects, birds and mammals when resting and during flapping flight. Redrawn from Bartholomew (1982).

the African savannahs that weighs about 13 kg. The take-off speed required to launch an ostrich would be in the region of 1000 km h^{-1}! Also, the largest flying mammal, the pteropid bat, weighs only 1 kg. At the other end of the spectrum the smallest bird, the humming-bird, weighing approximately 2 g, exhibits an energetic cost for flight which is merely an extrapolation of the energetic cost of flight for heterothermic moths (see Fig. 3.14). The large difference in metabolic scope between flying insects and birds is therefore due to the much lower resting metabolic rate of the moths and not to major differences in the energetic cost of flight in these two groups. Why then are there no flying insects larger than 30 g and no flying birds smaller than 2 g? Most physiologists would agree that if birds weighed less than 2 g the very high, mass-specific, metabolic rate could not be sustained with the essentially tidal respiratory system possessed by birds. The more efficient flow-through tracheal system of insects sustains flight to much lower body masses but, in spite of this respiratory efficiency, there are no large flying insects, although the fossil record has revealed the presence of giant dragonflies. There are no valid physiological reasons why flying insects should not reach the size of the largest flying birds and even go beyond this range. A really convincing answer has not yet been provided and there is probably a complex of ecological reasons for this phenomenon, including such factors as the avoidance of predation during larval development. These questions are clearly a subject for interesting speculation on the physics of flight, the physiology of respiration and countless ecological and evolutionary considerations.

Locomotion

To do justice to the infinite variety of locomotory mechanisms employed by animals would take volumes. One need only mention amoeboid pseudopodia, the cilia of paramecia, jet-propulsion by squid, the elegant flight of soaring raptors and the explosive muscle power of sprinting cheetahs to glimpse the vast field this subject covers. Alexander (1982) has written a lucid account of the major principles involved in the mechanisms of locomotion, their limitations and their energetic cost. We shall limit our discussion to energetic costs, but first some essential terminology.

Drag is produced by the viscosity of the medium through which the animal is travelling. For a given shape drag is proportional to surface area. Therefore large animals, because of their smaller relative surface area, experience less drag per unit mass. Because water is denser than air, greater drag forces are experienced in water but the flow pattern of the medium around the animal is also important. For example, if the change in velocity of the fluid is smooth as it moves progressively further from the surface of the animal, the flow at the boundary layer is laminar. In contrast, a steep gradient of velocity change causes eddy currents and secondary flow patterns and the flow becomes turbulent (Fig. 3.15).

Turbulence therefore lowers the efficiency with which metabolic energy is converted into propulsive movement. Consequently, the bodies of most fish and marine mammals have been shaped by natural selection to minimise drag and turbulent flow. The body surface must also offer resistance against the force of the flowing fluid to prevent the creation of localised secondary eddy currents. Obviously, drag is more important at high speeds of locomotion than when moving slowly.

Inertia is a function of body mass and small animals can accelerate more rapidly than large ones but due to their lower momentum (mass × velocity) they stop or decelerate more rapidly.

The *Reynolds number* is a dimensionless ratio and combines consideration of density and velocity of the medium through which an object is moving as well as the dimensions and velocity of the object. It is described by the following equation:

$$Re = \frac{pVL}{\mu}$$

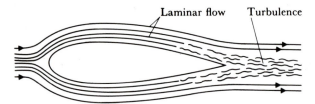

Figure 3.15 Laminar and turbulent flow of a fluid medium (gas or fluid) around a solid body. Laminar flow occurs when the pressure gradients between the surface of the body and the surrounding air layers are small.

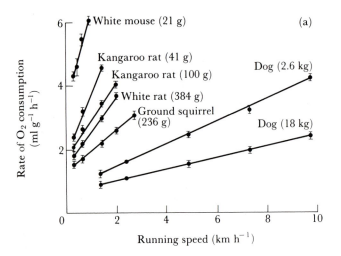

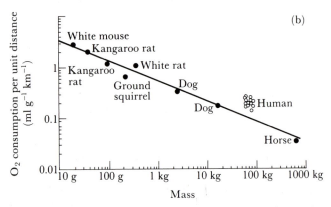

Figure 3.16 (a) Energetic cost of running (per unit mass transported per unit distance) plotted against running speed in different sized mammals. (b) Energetic cost of transporting 1 g 1 km in running mammals of various sizes. Redrawn from Taylor et al. (1970).

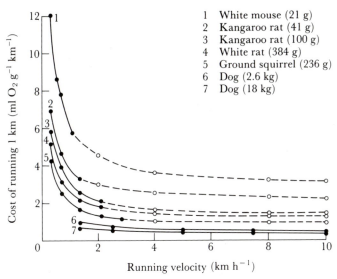

Figure 3.17 The cost of locomotion over a distance of 1 km is reduced with increasing running velocity until plateau values are reached. Dashed lines are extrapolations. Redrawn from Taylor *et al.* (1970) and Eckert and Randall (1983).

where p is the density of medium, V the velocity, L the appropriate linear dimension and μ the viscosity of medium. Actual examples or Reynolds numbers are: *Paramecium*, 0.001; *Daphnia*, 0.1; dragonfly, 1000; bird, 10^6; and whale 10^8. In other words, in very small animals viscosity of the medium dominates the ratio, whereas in large animals inertia dominates the equation. Eckert and Randall (1983) have in this regard compared the swimming of *Paramecium* in water as being the equivalent of a human swimming through warm tar. They also explain viscous drag by comparing the movement of a matchstick and a log when pushed along the surface of water at the same speed and then suddenly removing the propulsive force. The matchstick will come to an abrupt halt while the log will continue to coast. The same would be applicable to *Paramecium* and whales (Eckert and Randall, 1983).

The remainder of this discussion will consider the effect of speed and body size on the cost of locomotion as well as the comparative costs of running, swimming and flying. Figure 3.16b clearly shows that when the cost of locomotion is measured in terms of the cost of moving a unit of mass over a unit of distance, then large animals are much more effective than small animals. Figure 3.16a also informs us that the cost of locomotion increases linearly with increasing speed but that the rate of increase is much steeper in small animals. Before attempting to answer why large animals are more efficient locomotors, we should also note (see Fig. 3.17) that cost of loco-motion should also take the time, to move a certain mass over a certain

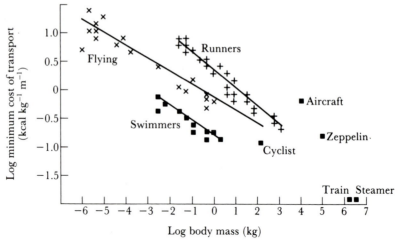

Figure 3.18 The relationship between log cost of transport and log body mass shows that swimming appears to be more efficient than flying and flying more efficient than running. Simplified from Tucker (1975).

distance, into consideration. In other words, even though an animal expends more energy per unit time running at high speeds, it reaches its destination sooner. The data in Fig. 3.17 show that although this cost is markedly reduced as velocity begins to increase it soon levels off at a fairly constant value in the case of animals of widely different sizes (Taylor *et al.*, 1970).

To explain why large animals are more efficient locomotors than small ones is not particularly easy. Richard Taylor of Harvard University is the leading international expert on the subject and he and his colleagues have advanced the following reasons for this phenomenon (Wingerson, 1983). Smaller animals experience greater drag than larger ones because of their greater relative surface area; this is, however, less important at low speeds. Because of their greater stride frequency smaller animals employ higher rates of muscle shortening to achieve a given velocity. The amount of stored elastic energy is less in small animals than large ones. This last reason is now widely accepted and can easily be appreciated by examining the mechanics of walking movements in a human. Starting at rest, the first requirement for movement is muscular contraction to move the centre of gravity of the body forward. Once the body is moving it can swing forward, without using very much energy as it will be exploiting gravitational forces much like a pendulum. The larger the person the greater the advantage of this pendulum action. The leg then moves forward and the knee locks to counter gravity. The force behind the impact of the foot on the substrate will not be entirely dissipated because a considerable amount of this energy (which includes free gravitational energy) will be stored as elastic energy in the tendons of the limb. This stored elastic energy can then be exploited on the next step when the tendons recoil and the limb is raised and moved forward. Again, the larger the animal the greater the amount of elastic energy that can be stored. Not all animals exploit this stored elastic energy as efficiently as running men and dogs do. The latter gain almost two-thirds of their forward power from this 'spring'. Lions do not exploit the spring-effect significantly, and although penguins are very efficient swimmers, when they waddle on land they are obviously not exploiting very much elastic recoil energy and are inefficient locomotors.

Figure 3.18 shows that swimming is the least expensive mode of locomotion, running is the most expensive and flying is intermediate. Aquatic animals experience a great deal of drag because of the high viscosity of water but buoyancy cancels out the forces of gravity, and streamlining of the body compensates considerably for the disadvantages of drag. Also, because of the advantages of buoyancy, aquatic animals can develop a greater potential power output from a large muscle mass which increases to the third power (l^3) while surface area becomes relatively smaller (l^2), thereby increasing the power/drag ratio.

When in flight, animals enjoy the advantages of moving through a low-density medium which reduces drag but they do not benefit from buoyancy. Forward propulsion through the air does, however, provide lift and a gliding bird is a very efficient locomotor. However, very small fliers are unable to glide because of high drag forces and must engage in flapping flight continuously.

Running animals are the least efficient because not only must the muscles counteract gravity to propel the body forward but they must also break the impact of the force when the animal makes contact again with the substrate. This is one of the reasons why walking downhill is comparatively costly and riding a bicycle is so efficient. Taylor (1987) has also pointed out that although a large terrestrial animal will locomote more efficiently than a small one, the mass-specific cost per stride (cycle frequency) at equivalent speeds is the same for large and small animals but the stride frequency is much greater in small animals (Table 3.5). He then advances a similar reason for the lower energetic costs for flying and swimming, namely that locomotion in water and air do not require as high a cycle frequency or mass-specific cost per cycle as during terrestrial locomotion.

To conclude, then, the cost of locomotion is of profound importance in shaping the life styles of animals and the practising ecologist should be well aware of the principles involved. For example, large animals are able to utilise far larger home ranges than small ones because they locomote more efficiently. The mass-specific metabolic rate of the larger animals is also less than that of the smaller animals and they are able to range over long distances in search of new sources of food and water. Nevertheless, there is a well-established inverse relationship between body size and abundance of animals within ecosystems. Why then are there far fewer large animals than

Table 3.5. The mass-specific energetic cost for each stride is the same for animals of different size at equivalent speeds where muscle stress (force/cross-sectional area) is the same. From Taylor (1987)

Body mass (kg)	Equivalent speed (m s^{-1})	Stride frequency at this speed (strides s^{-1})	Mass-specific cost/stride at this speed (ml O$_2$ stride^{-1} kg^{-1})
0.1	0.51	8.54	0.28
1.0	1.53	4.48	0.25
100	4.61	2.35	0.28

small ones within specific ecosystems? Because the absolute or total energy requirements of the large animals are much greater. Remember, however, that it is always wise to avoid over-emphasising a single environmental factor or physiological response when evaluating the life history or ecology of a species.

Evolution or natural selection frequently imposes compromises on animals, forcing them to undertake seemingly costly activities for small benefits, but closer examination will usually reveal that the activity is vital for survival. For example, the incredibly high cost of tunnelling into hard wood by certain carpenter bees for the sake of producing three or four offspring per year; or the enormous cost that honey-bees incur when air-lifting water to the hive for evaporative cooling (see Chapter 2).

Other major factors affecting energy requirements

Reproduction

This will obviously increase energy requirements in mammals because of the synthesis of new foetal tissue. In addition to this direct requirement, there is an increase in the work-load on the maternal organs such as the heart, liver and kidneys to produce and sustain the growth of the foetus. This amounts approximately to a 25% increase in RMR in mammals. For example, the Food and Agriculture Organisation (FAO) estimates an additional energy allowance of 40 000 kcal for a human pregnancy. Many mammals also deposit more fat in their tissues during pregnancy 'in anticipation' of the high energy demands of lactation. In non-mammalian species the cost of reproduction will vary enormously depending on the size and number of eggs produced, whether the offspring are altricial or precocious and the cost of nestbuilding and administering to and care of the young. The latter aspects will be discussed in Chapter 4.

Growth

This will naturally also require additional energy and the nature of the tissue which is being deposited will determine the actual energetic requirements for growth, e.g. adipose tissue has an energy content about 2.25 times greater than muscle tissue. Very refined estimates of the energetic cost of growth have been carried out on a few insects, laboratory rats and domestic animals. These data can be extrapolated to estimate energy budgets of whole populations.

Lactation

This appears at first to have a high energetic cost, but it is also the most efficient way of producing high-quality animal protein (70–80% efficient). This means that a human mother producing 850 ml milk per day, which contains 600 kcal, would only require 750 kcal per day extra.

The cost of *incubation* varies tremendously and depends on many factors, including environmental temperatures, the size of the bird, the number and size of the eggs, as well as the amount of energy the bird has invested in insulating the nest from the environment. Regression equations are available for estimating incubation costs.

General activity

Apart from routine locomotion, general activities will naturally increase the energetic cost of free existence. Grazing, chewing, foraging, territorial defence and courtship all represent increases over RMR and therefore have to be measured or estimated from previous studies when estimating the energetic cost of free existence. The following figures (from Moen, 1973) for a 100 kg ruminant are expressed as multiples of resting metabolism:

Basal metabolism	1.0
Standing	1.1
Running	8.0
Walking on level	1.64
Walking on 10% gradient	2.35
Foraging	1.59
Playing	3.0
Ruminating	1.26

3.2.6 Other methods for estimating energetic cost of free existence

Today, the DLW method of measuring FMR is accepted as the most precise estimate of an animal's total energy requirements for free existence (see section 3.2.3). However, the facilities for using this technique may not be available or the cost, particularly in large animals, may still be prohibitive. The investigator is then obliged to fall back on less precise methods of which the following are most commonly used.

Measurement of average daily metabolic rate (ADMR) in feeding experiments

The animal in question is placed in a fairly large cage or enclosure at ambient temperatures similar to its natural environment so that it can engage in reasonably normal daily activity, which approximates its natural behaviour in the wild. The amount and energy content of the food that is eaten is measured daily and if the animal does not gain or lose mass, this amount will represent the gross energy requirement for existence on that particular diet. A far superior experiment would include collection of the faeces and urine, as well as determinations of the energy content of these two items. In this way the metabolisable energy required for free existence can be derived:

Metabolisable energy = Gross energy − (energy in faeces + energy in urine)

Because the experimental animals are confined in fairly small cages, this technique obviously is of limited value and has been most successfully used on small rodents and certain insects.

Use of regression equations

The simplest and most reliable equations available are those published by Nagy (1987) and summarised in Table 3.4. These have been derived from scaling FMRs and their use has already been explained (section 3.2.3). Field metabolic rates are, however, not available for large ungulates and in the case of these animals it is still necessary to use the equations published by Moen (1973) in a most useful and instructive book entitled *Wildlife Ecology*. Moen's equations are based on values obtained from actual experiments on tame domestic animals (sheep and cattle) and can be used with reasonable confidence to estimate the cost of various life processes in similar animals in the wild, such as deer and antelope.

Time-based energy budgets

These have been most successfully applied to birds and humans because they can usually be kept under constant observation to record their activities continually. The procedure is fairly simple: the metabolic cost of various activities is measured using suitable apparatus. In the case of birds this will involve measuring oxygen consumption in a flow-through system while noting the activity exhibited by the bird. To obtain the cost of flying, the bird will have to be fitted with an oxygen mask and flown in a wind tunnel, or values obtained by other workers on similar species will have to be accepted. Humans can be fitted with a portable Max-Planck respirometer (back pack) which will allow measurement of oxygen consumption while they are engaged in their normal activities. Once the cost of the various activities has been established, all that remains is to keep the subject under close and continuous observation to estimate how much time is spent on each activity.

Table 3.6. **Energy budget of the male dickcissel. From Schartz and Zimmerman (1971)**

Activity	% day	Unit cost (kcal bird^{-1} day^{-1})	Total cost (kcal bird^{-1} day^{-1})
Night-time	37.50	17.2624	6.4734
Resting	6.50	17.2624	1.1220
Foraging	12.37	22.4411	2.7760
Singing	31.74	18.9886	6.0270
Courtship	0.61	29.3461	0.1790
Maintenance of female	4.91	29.3461	1.4409
Territory defence	0.27	34.5248	0.0932
Interspecific aggression	0.01	34.5248	0.0034
Distant flight	6.09	103.5744	6.3077
Total	100.000		24.4226

Table 3.7. **Energy expenditure by a coal-face miner. From Durnin and Passmore (1967)**

Activity	Time spent in one week hr	Time spent in one week min	Rate of energy expenditure (kcal min^{-1})	Energy in one week (kcal)
Sleep				
In bed	58	30	1.05	3 690
Non-occupational				
Sitting	38	37	1.59	3 680
Standing	2	16	1.8	250
Walking	15	0	4.9	4 410
Washing and dressing	5	3	3.3	1 000
Gardening	2	0	5.0	600
Cycling	2	25	6.6	960
Total	65	21		10 900
Occupational				
Sitting	15	9	1.68	1 530
Standing	2	6	1.8	230
Walking	6	43	6.7	2 700
Hewing	1	14	6.7	500
Timbering	6	51	5.7	2 340
Loading	12	6	6.3	4 570
Total	44	9		11 870
Grand total				26 460
Average daily energy expenditure				3 780

In this way an energy budget can be drawn up using a suitable unit of time. Examples of such budgets appear in Table 3.6 (bird) and Table 3.7 (human).

Table 3.8. **The gross energy value and nutrient content of some common foods (values per 100 g). From Durnin and Passmore (1967)**

	Energy (kcal)	Water (g)	Protein (g)	Fat (g)	Carbohydrate (g)	Alcohol (g)
Whole wheat flour	339	15.0	13.6	2.5	69.1	—
White bread	243	38.3	7.8	1.4	52.7	—
Rice, raw	361	11.7	6.2	1.0	86.8	—
Milk, fresh, whole	66	87.0	3.4	3.7	4.8	—
Butter	793	13.9	0.4	85.1	tr.	—
Cheese, Cheddar	425	37.0	25.4	34.5	tr.	—
Beef steak, fried	273	56.9	20.4	20.4	0.0	—
Haddock, fried	175	65.1	20.4	8.3	3.6	—
Potatoes, raw	70	80.0	2.5	tr.	15.9	—
Peas, canned	86	72.7	5.9	tr.	16.5	—
Cabbage, boiled	9	95.7	1.3	tr.	1.1	—
Orange with peel	27	64.8	0.6	tr.	6.4	—
Apple	47	84.1	0.3	tr.	12.2	—
White sugar	394	tr.	tr.	0.0	105.0	—
Beer,* bitter	31	96.7	0.2	tr.	2.2	3.1
Spirits* (gin, whisky 70% proof)	222	63.5	tr.	0.0	tr.	31.5

* Values per 100 ml.

Table 3.9. **Daily rates of energy expenditure by individuals with various occupations. From Durnin and Passmore (1967)**

Occupation	Energy expenditure (kcal day^{-1})		
	Mean	Minimum	Maximum
Men			
Elderly retired	2330	1750	2810
Office workers	2520	1820	3270
Colliery clerks	2800	2330	3290
Laboratory technicians	2840	2240	3820
Elderly industrial workers	2840	2180	3710
University students	2930	2270	4410
Building workers	3000	2440	3730
Steel workers	3280	2600	3960
Army cadets	3490	2990	4100
Elderly peasants (Swiss)	3530	2210	5000
Farmers	3550	2450	4670
Coal miners	3660	2970	4560
Forestry workers	3670	2860	4600

Remember that additional estimates for growth, temperature regulation and reproduction must be included if the energy budget for a whole population is being calculated on a long-term basis. Also care must be taken to avoid the common mistake of using gross energy values of foodstuffs to calculate how much of a particular food is required to meet the requirements of the energy budget. Obviously the amount of metabolisable energy or, if at all possible, the net energy of the food must be used. Ecologists are often required to give advice about management decisions involving large ecosystems and although these energy budgets are not by any means exact, they are infinitely preferable to guessing in the dark and can be very valuable for analysing ecological problems particularly management of large game reserves.

For the sake of interest the *gross* energy values of some common foods are presented in Table 3.8 and the daily rates of energy expenditure in various occupations in Table 3.9. You can amuse yourself by calculating balanced diets for male army cadets for a year based on this and other information.

3.2.7 The reduction of energy requirements

When endotherms are faced with periodic nutritional and/or temperature stress, it would obviously be an advantage to reduce energy requirements by lowering the metabolic rate and abandoning homeothermy. This state of hypothermia and greatly reduced metabolic rate is known as torpor if it occurs on a diurnal basis or hibernation if it occurs seasonally. Torpor should not, however, be confused with hypothermia, which is a prelude to death, unless an external source of heat is applied. In true torpor or hibernation the animal is capable of spontaneous arousal without the application of external heat and is therefore quite different to both hypothermia in endotherms and the poikilothermy of ectotherms.

Small birds and mammals, because of their high relative surface areas and inability to store large amounts of nutrients, are obvious candidates for either torpor or hibernation. It is particularly prevalent in rodents and insectivores while many orders of birds exhibit torpor and at least one of the nightjars (Caprimulgidae) shows true seasonal torpor. The dramatic reduction in metabolic rate accompanying diurnal torpor in the little brown bat and nocturnal torpor in Allen's humming-bird is illustrated in Fig. 3.19. Larger mammals, such as badgers, racoons, skunks and bears experience a so-called 'winter lethargy' during which metabolic rates and body temperatures are very significantly reduced, but not nearly to the same extent as in small mammals. For this reason the latter are sometimes referred to as true hibernators, although the differences involved are largely a matter of degree. A distinction is also made between seasonal or obligate hibernators and facultative hibernators. In the case of the former the condition is seasonally controlled by a circannual, endogenous rhythm which is entrained long before the beginning of winter stress. It is a spontaneous phenomenon, starts slowly and occurs regularly, e.g. Arctic ground squirrel. Facultative hibernators, on the other hand, can be induced to hibernate by merely removing food

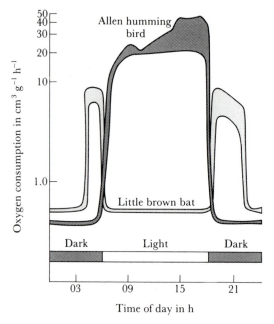

Figure 3.19   The fluctuation in metabolic rate of the Allen humming-bird and the little brown bat during periods of torpor and activity. Redrawn from Richards (1973).

or subjecting the animals to cold stress. The North American pocket mouse is an example of such an animal and hibernation in these animals is opportunistic, has a rapid onset and is irregular.

Entry into hibernation is not yet completely understood in spite of the voluminous literature on this subject. Environmental cues such as the photoperiod and ambient temperature appear to be important in entraining vital changes in the neuroendocrine system. Hibernation itself is accompanied by both morphological and functional involution of all the endocrine glands other than the beta cells of the pancreas. Entry into hibernation is also often accompanied by test drops in body temperature, heart rate and oxygen consumption with periodic returns to homeothermy before full hibernation occurs.

The state of hibernation is not one of continuous deep sleep. Sleep, particularly during the early phases, can be interrupted by periods of arousal when the animal may engage in various activities such as eating, drinking and the elimination of faeces and urine. The body temperature of the hibernator is usually close to that of the ambient air, but if temperatures fall too low, metabolic rate will increase to protect the tissues from freezing. Endothermy has therefore not been completely abandoned although metabolic rates have fallen as low as 1–3% of the normal rate. Respiratory quotients are always in

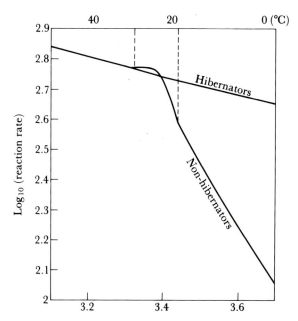

Figure 3.20 A comparison of enzyme reaction rate of hibernators and non-hibernators at different temperatures.

the region of 0.7, indicating heavy dependence on lipid catabolism to supply energy for the various vital processes. Catabolism of lipid also yields large amounts of metabolic water which is an obvious advantage to the hibernator. The volume of circulating blood is reduced because of blood storage in the spleen and liver but the slow-moving blood is in no danger of coagulating as the clotting time doubles during hibernation. The breathing patterns of hibernating animals become very irregular because of the much lower oxygen demand and the reduced CO_2 production. Nevertheless, those resting in closed dens or burrows must at times cope with ambient air containing as much as 6.2% CO_2 and as little as 13.7% O_2.

Hadley (1985) has reviewed adaptations in membrane function which hibernators exhibit. He points out that non-hibernators cannot tolerate long periods of hypothermia and if forced to do so there is a disruption in membrane enzyme activity. This is illustrated in Fig. 3.20 which shows discontinuities of the Arrhenius plot at a temperature around 20 °C. These discontinuities are considered to be due to phase shifts of membrane lipids. Hibernators do not exhibit this phenomenon and enzyme activity follows a straight-line relationship (Fig. 3.20). The adaptations in membrane function which protect hibernators are not completely understood but Hadley (1985) suggests the following possibilities: enhancement of unsaturated fatty acids in the membranes, which reduces the temperature at which the membrane shifts from a liquid crystalline state to a gel-like state; insertion of cholesterol

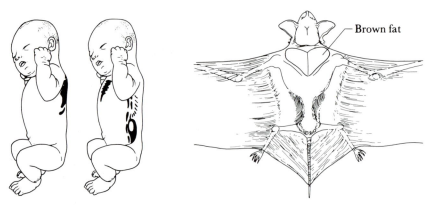

Figure 3.21 Distribution of BAT in the neonate human and bat. Redrawn from Hadley (1985).

as a spacer molecule; and augmentation of phospholipids containing weakly bonding polar groups. These adjustments may be very localised to a specific region of the membrane.

Arousal from hibernation is usually a rapid but costly process. Richards (1973) reports that some 90% of the total heat production during hibernation occurs during arousal and the remaining 10% during the entire period of torpor. This emphasises the very real savings in energy expenditure that hibernation can achieve. The cause of the rapid arousal from hibernation can be attributed almost solely to the impressive thermogenic potential of BAT, which surrounds the vital internal organs and great vessels of these small mammals. The temperature of the heart and BAT rises first and there is a blood shunt to the thoracic muscles. The RQ is predictably close to 0.7 and there is non-shivering thermogenesis. Metabolic rate rises so rapidly that a temporary 'overshoot' occurs. Brown adipose tissue is therefore essential for so-called true hibernators and is of sufficient physiological and ecological interest to warrant further attention.

3.2.8 Brown adipose tissue

Because BAT occurs in all true hibernators it was once known as the hibernating gland. It has since been shown that BAT does not secrete specific chemicals into the general circulation and the term has been abandoned. Brown adipose tissue also occurs in many non-hibernating, young mammals, including human infants. Adult small mammals frequently employ BAT for non-shivering thermogenesis and the amount of this tissue usually increases when the animals are exposed to cold. Typical locations for the deposition of BAT are between the shoulder blades, around the neck, heart, great vessels and lungs. These strategic positions ensure rapid warming and normal func-

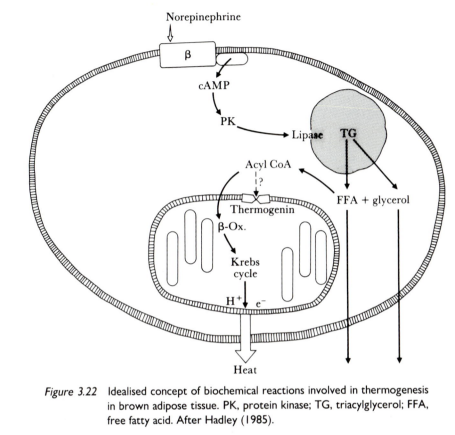

Norepinephrine

β

cAMP

PK

Lipase TG

Acyl CoA

?

Thermogenin

FFA + glycerol

β-Ox.

Krebs
cycle

H⁺ e⁻

Heat

Figure 3.22   Idealised concept of biochemical reactions involved in thermogenesis
in brown adipose tissue. PK, protein kinase; TG, triacylglycerol; FFA,
free fatty acid. After Hadley (1985).

tioning of the vital organs during arousal or protection of these organs during
cold exposure (Fig. 3.21).

The characteristic brown colour of this tissue is due to both its rich
vascularisation and the inclusion of large amounts of cytochrome, associated
with a very high abundance of mitochondria. The large number of mitochondria make it possible for this tissue to oxidise fatty acids very rapidly and
thereby transform chemical energy into heat. Dawkins and Hull (1965) have
shown that BAT in rabbits can oxidise succinic acid 20 times more rapidly
than white fat tissue. The speed of oxidation is, however, not the only
characteristic of this tissue which allows it to produce such large amounts of
heat. It would also seem that synthesis of ATP in BAT can be reduced, while
heat production is concomitantly increased. This is apparently due to the
presence of a proton conductance pathway on the inner mitochondrial membrane. The sequence of events can be summarised as follows. Thermogenesis
is primarily activated by the sympathetic nervous system through the release
of noradrenaline. The noradrenaline binds to β-adrenergic receptors, which
can number as many as 150 000 per adipocyte. This stimulus results in

increased synthesis of cyclic AMP, which then acts as a second messenger causing the activation of a specific protein kinase. The protein kinase activates lipase, which hydrolyses the triacylglycerols producing glycerol and free fatty acids in the normal way (Fig. 3.22). The glycerol cannot be metabolised within the BAT and is transported to the liver and muscle tissue for further processing. Acyl-CoA enters the mitochondrion where it is exposed to oxidation and the enzymes of the Krebs cycle to eventually produce CO_2 and H_2O. Normally at this stage protons (H^+) are pumped out of the mitochondria to provide a steep proton electrochemical gradient which is essential for ATP synthesis. It is now thought that in BAT the mitochondrial membrane becomes more permeable to protons, thereby dissipating the proton electrochemical gradient, reducing ATP synthesis and increasing heat production (Fig. 3.22). In addition, it should also be noted that the mitochondria in BAT possess a specific protein, thermogenin, which controls the opening of channels in the mitochondrial membrane for the oxidation of fatty acids without the normal levels of ATP synthesis. The subject of BAT has been fully reviewed by Hadley (1985) and the above description has made liberal use of his ideas.

3.3 Case studies

3.3.1 Marine invertebrates

The energy metabolism and nutrition of marine invertebrates, particularly those inhabiting the inter-tidal zone, have received considerable attention but probably the most bizarre examples are provided by marine invertebrates living close to deep-sea hydrothermal vents. The ability of these animals to survive at such great depths and, consequently, tremendous pressures is sufficient cause for astonishment, but when one considers that they also have to cope with anoxic or at least hypoxic water which often has sulphide levels well above toxic thresholds, then they are obvious candidates for study. Basically, the nutritional problem facing certain marine invertebrates at these hydrothermal vents is to avoid the toxicity of the sulphides, but at the same time supply the endosymbiotic bacteria in their tissues with sulphide for their metabolic requirements. To complicate matters further the endosymbiotic bacteria that are able to oxidise sulphides to produce energy are also sensitive to excessively high levels of sulphide, especially in the presence of oxygen. The physiological mechanisms employed by the animals must therefore protect their own tissues as well as the endosymbionts from the toxic effects of high levels of oxygen and sulphide, while simultaneously supplying the symbionts with sufficient sulphide. Childress (1987) has written a short review on the subject and the following description of the mechanisms employed by the vestimentiferan tube worm *Riftia pachyptila* will be our example and is taken from his review.

Riftia pachyptila has the same basic morphology as other pogonophorans and lives in the immediate vicinity of deep-sea hydrothermal vents where sulphide-rich anoxic water from the vents is mixing with sulphide-free water containing oxygen. The interior of this tube worm's trunk is mostly occupied by the trophosome which also contains the bacterial symbionts. The major organ of gas exchange is the plume and there is a vascular connection between the trophosome and the plume. The coelomic fluid and the blood within the vascular system each contain their own type of haemoglobin and it appears as if there is little or no mixing of these two fluid compartments. Both haemoglobins have a high affinity for oxygen and allow the provision of relatively high concentrations of oxygen to the symbionts even when oxygen concentration of the ambient water is low.

The blood of freshly captured *R. pachyptila* contains 10 times the sulphide concentration of the surrounding water and this concentrating ability is attributed by Arp and Childress (1983) to reversible binding of sulphide by the haemoglobins at a non-haem site. Spontaneous oxidation of the sulphide is retarded by low partial pressures of oxygen and by keeping concentrations of sulphide and oxygen sufficiently low. It is also important to note that the toxic effect of hydrogen sulphide is similar to that of cyanide, namely inhibition of cytochrome-*c*-oxidase, but that the affinity of *R. pachyptila*'s haemoglobin for sulphide is greater than that of its cytochrome-*c*-oxidase. For this 'safety mechanism' to operate effectively, however, the haemoglobin should never become saturated with sulphide. This is ensured by the removal of sulphide from the haemoglobin by the endosymbiotic bacteria. They, in turn, oxidise the sulphide to produce energy. They may only oxidise the sulphide to elemental sulphur if oxygen is limited. The elemental sulphur can then serve as a non-toxic energy store which can be oxidised further when required.

It should be emphasised that the preceding description represents only one mechanism and one species. Several different mechanisms have been discovered in various species among the bizarre deep-sea community. For a recent review on this subject, see Childress (1987).

3.3.2 Arthropods

Nutritional studies on arthropods are legion and it is impossible to do justice to this subject within the limited length of this review. It is hoped, however, that this short selection on nutritional and energetic studies will at least serve to whet the appetite for more extensive reading among the references.

As we have already noted, many plants mount complex chemical defences against herbivorous insects by synthesising so-called secondary compounds, while the insects have evolved an array of adaptations to deal with these defences. This 'arms race' between the plants and herbivores has been in progress for millions of years and, if one looks around at the mass of green herbage covering most of the land masses, then clearly it would seem as if the plants are still winning the race. Nevertheless, the counterploys evolved by insects are very effective as evidenced, for example, by the vein-cutting behaviour of certain insects before feeding, as described by Dussourd and Eisner (1987). These authors performed a series of simple but imaginative experiments to show that a number of insect species, from different phyletic lineages, that feed on latex-producing plants will first cut the veins of the leaf before feeding distal to the cut. The cutting of the leaf vein blocks latex flow to the feeding sites and allows the insect to feed unhindered.

These authors also studied the feeding behaviour of the chrysomelid beetle *Labidomera clivicollis* upon field milkweed, *Asclepias syriaca*. They provided this beetle with the choice of feeding on an undamaged leaf or one in which the veins had been cut on one side of the midrib artificially. *Labidomera* definitely preferred the half of the experimental leaf which had been damaged. In other words, they prefer leaf tissue that has already been damaged to leaves which they must first process themselves. They also established that generalist feeders, that do not practise vein cutting, will feed on leaves prepared by vein-cutting species who are specialist feeders. Vein cutting is not restricted to laticiferous plants. A most interesting example of leaf trenching by *Epilachna* beetles on squash (cucurbit) leaves has been described by Tallamy (1985). The act of leaf trenching apparently blocks the translocation of bitter-tasting cucurbitacins beyond the site of damage. Normally these deterrent chemicals can be mobilised and translocated by the plants within 40 min of insect damage.

In spite of the very many studies on insect nutrition, few have attempted to examine the energy budgets of insects under natural conditions. Perhaps the best studied species in this regard is the bumblebee (*Bombus* spp.) and Bernd Heinrich has written an intriguing account of these studies entitled *Bumblebee Economics*. They are too extensive to describe here and instead we shall discuss a restricted study on the energy and water balances of bumblebees by Bertsch (1984) and one on the energetic cost of reproduction in carpenter bees (Louw and Nicolson, 1983).

Bertsch (1984) studied male bumblebees confined in a large climatic chamber provided with artificial flower feeders. The latter were designed to allow the measurement of the amount of nectar taken by the bees per 'flower visit'. Flight activity was monitored, the bees were weighed regularly and in a separate experiment, O_2, CO_2 and H_2O exchange were measured while the bees were at rest and while flying. The mean total time spent flying was 244 min over a mean distance of 17 km. While flying the bees consumed 56.4 ml $O_2 g^{-1} h^{-1}$ and lost 6 mg $H_2O g^{-1} h^{-1}$. The bees also consumed an amount of nectar (180 µl 50% sugar solution) equal to their own body mass each day. The body mass remained constant and approximately 70% of the energy was used for the 244 min of flying time per day. At the same time the bees produced 66 mg of metabolic water, consumed 110 mg of water in the nectar solution and lost 40 mg of water by evaporation. This means that 136 mg of water must have been excreted daily from the digestive tract for the bees to remain in water balance. This amount equals the total body water of an average bumblebee and clearly shows that nectar provides excessive amounts of water, which add to the payload of the bees and consequently to their flying costs. It is not surprising therefore that they must, like carpenter bees (Nicolson and Louw, 1982), produce copious volumes of a dilute urine. The same considerations naturally also apply to the energy and water balances of nectar-feeding bats and birds although the larger body size would mean a lower cost of locomotion and therefore a lower production of metabolic water. Birds and bats would also lose more water by evaporation per unit O_2 consumed but still face real energetic stress when nectar concentrations become very dilute. In this regard it is of interest to note that Bertsch (1984) employed a nectar concentration of 50% sugar, whereas Baker and Baker (1982) report that the mean sugar concentration of bee nectar is about 35% in northern temperate areas. Bertsch (1984) also reports that the male bumblebees in his experiments spent 60% of the daily light period roosting, in spite of the unlimited supply of energy available in the artificial flowers. He advances this fact as an argument against the conventional view of 'optimal foraging' and suggests that water-loading is the limiting factor in this species and not energy.

There are no bumblebees in Africa south of the Sahara and in some ways carpenter bees, which provide our next example, appear to fill this vacant niche. A particular species *Xylocopa capitata*, because it feeds almost exclusively on one species of plant (*Virgilia* sp.), provides ideal material for a field study of the energetic cost of free existence and for reproduction. Such a study has been undertaken by Louw and Nicolson (1983) in the Fynbos vegetation at the southern tip of Africa. They first flew bees in specially constructed respirometers and found that the cost of free flight was equivalent to the consumption of about $52 \, ml \, O_2 \, g^{-1} h^{-1}$, which for a 1.5 g bee represents an energy expenditure of $1570 \, J \, h^{-1}$. The average foraging female has a body temperature of 39 °C and carries 68.5 µl of sugar solution in her crop at a concentration of $75 \, g \, 100 \, ml^{-1}$. This is equivalent to an energy store of 903 J and is sufficient fuel for 34.5 min of continuous flight. *Virgilia* flowers, on the other hand, contain 0.7 µl of nectar on average with a sugar concentration of $66 \, g \, 100 \, ml^{-1}$; thus a single flower provides a small packet of energy equal to 8.1 J. Therefore to maintain itself in near continuous flight without diminishing the sugar stores in its crop, a bee must obtain the nectar from 194 *Virgilia* flowers per hour or 3.2 flowers min^{-1}. Independent field observations, however, showed that only 67% of foraging time was spent in actual flight and foraging *X. capitata* visited 8 flowers min^{-1}.

Xylocopa capitata bore tunnels as long as 25 cm into the hard wood of dead trees in which they construct cells for their offspring. Each cell is provisioned with a round ball of pollen paste consisting of a mixture of pollen and nectar, upon which a very large single egg (c. 123 mg) is laid. Only one batch of eggs ($\bar{x}$, 4) is laid in the lifetime of the female, obviously a strongly k-selected species. The energetics of reproducing in this way was studied by determining the energy value of fully grown larvae, faeces and complete pollen balls recovered from intact nests. These studies were greatly facilitated by the fact that complete development occurs within an individual cell, so that it was possible to collect the total food available and total faeces produced, as well as the animals in various stages of the life cycle. The mean energy provided by a single pollen/nectar ball was 43.01 kJ and during the course of development the average larvae excretes 4.38 kJ of egesta, loses 15.41 kJ in respiration while depositing 23.22 kJ of energy in its tissues. This gives a mean assimilation efficiency for the larvae of 89.8% and a gross production efficiency (production/consumption) of 54.0%. Brafield and Llewellyn (1982) have reviewed the gross production efficiencies of a wide variety of ectotherms and these vary from 6% (terrestrial detrivore) through 24% (terrestrial granivore) to 42% (certain parasites). Larval growth in the carpenter bee, at an efficiency of 54%, is in the highest range and is largely due to the high quality of the food used to provision the cells and minimal activity required of the larvae.

Louw and Nicolson (1983) further calculated that the average female X. *capitata* must visit about 1678 flowers to provide sufficient pollen to raise one adult. At a flower visitation rate of 8 flowers min^{-1}, it would require a total period of 14 hours in actual foraging time to raise the maximum number of four adults per annum. This is a surprisingly short time and one can only conclude that the very low reproductive rate of this species is not the result of the high cost of foraging but due to some other 'high-cost' energetic demand in their lives. The most obvious of these would appear to be the cost of constructing tunnels in hard, dead wood. In this way, although very few offspring are produced, they are very well protected within the hard wood tunnels and the benefits must obviously outweigh the high cost of tunnelling. To conclude, we should also note that this species of carpenter bee, because of its diet and high metabolic rate, experiences the same problem of water loading as the bumblebee. Consequently it produces copious amounts of a dilute urine (155 mosmol kg^{-1}),

The competition among researchers to develop the most sensitive and accurate means of measuring gas exchange in small animals was bound eventually to face the ultimate challenge of measuring oxygen consumption in ants. This has now been achieved by Bartholomew *et al.* (1988) at the University of California laboratories in Los Angeles and at the Smithsonian Tropical Research Institute in Panama, where they examined the energetics of trail running, load carriage and emigration in the column-raiding army ant *Eciton hamatum*. Army ants are nomadic and the bivouac of the colony is moved each day for distances as much as 90–100 m. During these movements the larvae are carried from one bivouac to the next. From time to time the ants enter the statory phase when the bivouac remains in one place while the pupae mature and a new brood of eggs is laid. During the latter phase raiding parties leave the colony and return with food. Therefore the carrying of either brood or food is an essential and everyday occurrence in the life style of these ants and an important part of their energy budget. Interestingly, an army ant lifts its load with its mandibles while straddling the load and carries it slung beneath the body. To measure the cost of load carriage Bartholomew *et al.* (1988) were obliged to use living pupae as the test loads. For this reason they had to first determine the oxygen consumption of various sized pupae in order to subtract this amount from the total oxygen consumption of the load-bearing ant plus pupa. They found that the gross cost of transport and the net cost of transport decreased with increasing body mass and running speed. The mean rate of oxygen consumption of load-carrying ants (mean mass, 10.8 mg) was $5.72 \, \text{ml} \, O_2 \, g^{-1} \, h^{-1}$ and the RQ suggested that lipids were mainly being catabolised to supply the required energy. Compare the above oxygen consumption with that of the cost incurred by a honey-bee transporting water to the hive ($c. 95 \, \text{ml} \, O_2 \, g^{-1} \, h^{-1}$).

Bartholomew *et al.* (1988) have also used their data to develop a model in the form of a computer program to predict the energy spent for maintaining a raiding column, for colony emigration and for a colony while in bivouac. They also make an interesting comparison of field energetics in army ants and in leaf-cutting ants. Their findings show that the army ants, although smaller, are able to run faster while carrying loads that are lighter in relation to their body mass. The leaf-cutter ants, in contrast, are obliged to carry bulky and asymmetrical leaf and flower fragments, which are difficult to balance and can offer resistance to wind. These loads can be as much as 10 times their body mass, with the further disadvantage that the percentage of metabolisable energy in their food is lower than in the more concentrated food source transported by the army ants. These results are in keeping with our perception of the highly active, mobile and nomadic life style of army ants. Intuitively one would have guessed that locomotion and load-carrying would be relatively efficient in this species.

A novel case study of the energetics of locomotion in arthropods is provided by the cartwheeling, white lady spider of the Namib Desert dunes. Discovery of the wheel by man, together with the discovery of fire, are considered by many archaeologists as the major technological advances in the history of mankind, eclipsing even the discovery of the microprocessor in some eyes. Be that as it may, the superior efficiency of wheeled locomotion is indisputable but it is almost absent in the animal kingdom. Granted that the flagellae of certain unicellular organisms exhibit a circular whiplash action and that the cilia of rotifers appear to be wheeling these tiny creatures along, nevertheless, to the best of my knowledge, the only true example of wheeled locomotion occurs in the Namib Desert spiders (*Carparachne* spp.). These fairly large spiders are mostly nocturnal and leave their nests to hunt various

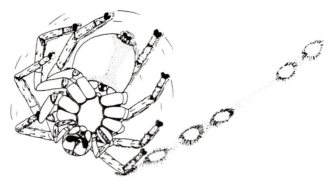

Figure 3.23   The posture of the dune spider, *Carparachne aureoflava*, as it cart-wheels down the dune slope from right to left. From Henschel (1990).

invertebrates on the surface of the dunes. Should these aggressive hunters, however, be disturbed and chased by a predator, they will run swiftly to-wards a suitable downward-sloping face on the dunes and roll down the slope with their appendages extended in the form of a cartwheel. Henschel (1990) has studied the phenomenon in some detail and reports that speeds between 0.5 and $1.5 \, \mathrm{m \, s^{-1}}$ are attained by the spiders and that they control the speed by progressively extending their appendages to full reach, much as ice skaters can control the speed of a spin by extending their arms. The reason why they employ this novel locomotion is not entirely clear. Obviously it is a rapid and energetically efficient form of escape but the spiders are in any event swift locomotors. Perhaps the complete change of locomotory pattern startles and confuses the diurnal predator, pompilid wasps, which perhaps have a programmed 'search image' of the spider (Fig. 3.23). Alternatively, spiders do not have well-developed aerobic scope and cartwheeling may allow the spiders to continue their escape in spite of having exhausted their oxygen reserves. They could then rest at the bottom of the slope and repay their oxygen debt before ascending the dune slope once again.

3.3.3 Amphibians

Relatively few studies have been carried out on the nutrition and energetics of amphibians, when compared with other vertebrates. To the best of my knowledge there has also not been any published work on the FMRs of amphibians using DLW.

Standard measurements of oxygen consumption under laboratory con-ditions show considerable interspecific variation. Values of $22.5–25.6 \, \mathrm{mm^3}$ $O_2 \, \mathrm{cm^{-2} \, h^{-1}}$ have been reported for various species of anuran, whereas Withers *et al.* (1982) recorded lower values for three species of African tree frogs (*Hyperolius* spp.), namely $10.5–11.1 \, \mathrm{mm^3 \, O_2 \, cm^{-2} \, h^{-1}}$. Steele and Louw (1988) report that the rare caecilian *Scolecomorphus kirki*, although exhibiting

Table 3.10. Classification of life styles in anurans in relation to the aerobic dependence index (ADI). Predator avoidance behaviour: A, active; S, static. Habitat: A, arboreal; T, terrestrial; SA, semiaquatic. Mode of locomotion: H, hop; J, jump; W, walk; B, burrow. Predatory behaviour: A, active searcher; P, passive searcher. From Taigen et al. (1982)

Species	ADI	Predator avoidance behaviour	Habitat	Mode of locomotion	Predatory behaviour
B. americanus	1.96	S	T	H-B	A
B. calamita	1.74	S	T	W-B	A
G. carolinensis	1.19	A	T	W-H	A
K. pulchra	1.03	S	T	W-H-B	A
H. crucifer	0.97	S	A	J	P
K. weali	0.96	S	A	W	—
O. americanus	0.69	S	T	H-B	—
P. fodiens	0.09	S	T	H-B	A
B. orientalis	−0.06	S	SA	H	—
K. senegalensis	−0.30	A?	T	W	P
A. callidryas	−0.45	S	A	J-W	P
R. sylvatica	−0.72	A	T	J	P
H. arenicolor	−0.95	A	A	J	—
H. viridiflavus	−0.98	A	A	J	—
O. septentrionalis	−1.20	—	A	J	P
D. pictus	−1.90	A	SA	J	P
E. coqui	−2.10	S	A	J-W	P

high evaporative water loss rates and employing cutaneous respiration, has a low resting metabolic rate (0.052 ml O_2 g^{-1} h^{-1}), which lies within the range established for plethodontid salamanders by Whitford (1973).

Taigen et al. (1982) in an extensive and imaginative study have measured aerobic and anaerobic metabolism during maximum exercise in 17 species representing seven amphibian families and a variety of ecological types and locomotor modes. In this study body mass ranged from 0.9 to 30.7 g. Resting metabolic rates ranged from 0.026 to 0.111 cm^3 O_2 g^{-1} h^{-1} with a mean and standard deviation of 0.061 ± 0.024 cm^3 O_2 g^{-1} h^{-1}. The maximum rates of oxygen consumption ranged from 0.266 to 1.043 cm^3 O_2 g^{-1} h^{-1} with a grand mean and standard deviation of 0.668 ± 0.226 cm^3 O_2 g^{-1} h^{-1}. Lactate content of the muscles after maximum exercise served as the measure of anaerobic metabolism and these values showed similar high variation ranging from 0.329 to 1.285 mg lactate g^{-1} tissue. After normalising these data, Taigen et al. (1982) calculated an aerobic dependence index (ADI) by subtracting the normalised lactate values from the normalised values obtained for maximum oxygen consumption for each species. A high ADI value indicates a strong dependence on aerobic energy production, whereas values below zero indicate much anaerobic capability. The ADI values thus obtained for each species are presented in Table 3.10 and compared with an

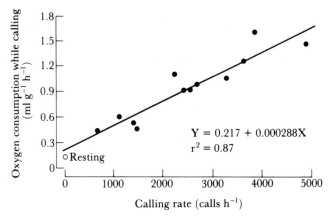

Figure 3.24 Metabolic cost of calling plotted against calling rate in male spring peepers. Note resting metabolic rate denoted as ○. Redrawn from Taigen *et al.* (1985).

assessment of various behavioural characteristics of the species. Taigen and his colleagues conclude from these comparisons that the diversity of metabolic patterns among anurans is great and is closely associated with many features of a particular species' biology. For example, foraging behaviour (active versus passive searching) and mode of locomotion (non-jumpers versus jumpers) were correlated with a species' dependence upon aerobic metabolism. Species with low aerobic capacity tend to make fewer capture attempts than those with higher aerobic scope. They also move over shorter distances in search of their prey and these results suggest that opportunistic switching of foraging behaviour is probably not possible because of metabolic constraints. Another example is provided by a species of toad that does not undergo dispersal until its aerobic capacity has developed to a threshold level (Taigen and Pough, 1985). Non-jumping anurans have a low metabolic performance, they cannot escape predators with a sudden leap and therefore usually rely on other defence mechanisms such as the secretion of toxic exudates. Success in combat between males of these toxic species may be favoured by a high ADI value. These authors caution, however, against considering single physiological and ecological features in isolation and commend an integrated approach when evaluating the physiological ecology of species.

Calling in male frogs has long been a subject for intensive study by biologists. Few, however, realise that this behaviour can be very costly in terms of its energetic cost and should therefore be included in energy budgets and general assessments of an anuran's physiological ecology. Taigen *et al.* (1985) have estimated the cost of calling in male spring peepers (*Hyla cruciferi*) together with various enzyme concentrations of the muscles involved in vocalisation of this species. They found that calling rate was strongly correlated with oxygen consumption (Fig. 3.24). The peak value for oxygen consumption during calling was 1.7 ml O_2 g^{-1} h^{-1}, whereas the peak value produced by vigorous exercise was only 1.1 ml O_2 g^{-1} h^{-1}. The trunk muscles are the major muscles involved in calling and these represent 15% of the total body mass of males and only 3% of the body mass of females. Citrate synthase activity and β-hydroxyacyl-CoA dehydrogenase activity were approximately five times greater in the trunk muscles than in the leg muscles indicating an unusually high oxidative capacity of these muscles. In fact, Taigen et al. (1985) conclude that the aerobic scope of these trunk muscles is the highest yet reported for ectothermic vertebrate muscle and that it even approximates the aerobic scope of endothermic muscle with high oxidative capacity.

3.3.4 Reptiles

Many reptiles are reasonably easy to maintain in captivity and often equally easy to observe under natural conditions in the wild. For these reasons and also because of their interesting ectothermic temperature regulation, a great number of field and laboratory studies have been carried out on reptiles. This is reflected in the appearance of a volume entirely devoted to reptiles, *Ecophysiology of Desert Reptiles*, by Bradshaw (1986). We are, however, obliged to limit our discussion here to a few key examples.

The nutritional requirements *per se* of lizards have not been extensively studied but field observations on general feeding habits are commonplace. These studies have shown that fewer than 2% of extant lizards are predominantly herbivorous and most of these species tend to have a body mass greater than 100 g. Pietruszka et al. (1986) have studied the predominantly herbivorous diet of the sand dune lizard *Angolosaurus skoogi*. These unusual lizards live on and in the barchan dunes on the extremely arid and desolate coasts of northern Namibia and southern Angola. Analyses of the faecal pellets of these animals revealed that 18 food taxa were being consumed: 81% by mass was composed of vegetable matter, of which 56% was an endemic leafless, but very thorny cucurbit, *Acanthosicyos horrida*. Seeds were also a significant component in the diet and it is of importance to note that the seeds were first husked by cracking them between the teeth before they were swallowed. These husking bouts could be heard from several metres and undoubtedly served to enhance the digestibility of the seeds. The major food plant *A. horrida* not only provided food in the form of flowers, shoots, fruit and green thorns, but also served as an important source of water. The plants have deep roots and in spite of the arid environment they are always under positive turgor pressure. The animals were frequently observed to bite off the tips of emergent stems, whereupon plant sap would exude and bead on the severed tip and be consumed immediately by the lizards. It is therefore not surprising that the distribution of *Angolosaurus skoogi* follows the distribution pattern of *A. horrida* very closely along this hyper-arid coast. The puzzle that remains, however, is to explain how the newly emergent seedlings of *A. horrida* survive until their tap-roots reach the moist sand far below the base of the sand dunes.

Anaerobic metabolism is a prominent feature in the energy metabolism of reptiles and Pough and Andrews (1985) have examined the use of anaerobic metabolism by free-ranging lizards. They carried out their studies in Cave Creek Canyon in Arizona at an altitude of 1650 m. They recorded various kinds of behaviour in the field before killing the lizards by shooting them with dust shot and placing them in liquid nitrogen for lactate analyses. They studied two species of teiid lizards *Cnemidophorus exsanguis* and *C. sonorae* and two species of iguanids *Sceloporus virgatus* and *S. jarrovi*. Lactate concentrations were lowest in the morning when the lizards first emerged. The teiids exhibited significantly higher lactate levels during emergence and after routine activity than did the iguanids. Capturing, subduing and swallowing grasshoppers resulted in a 40% increase in the amount of lactate produced when compared with levels recorded for routine activity. Territorial combat caused a 100% increase in lactate levels over those produced by routine activity. Pough and Andrews therefore understandably conclude that anaerobic metabolism can be an important pathway for providing energy to lizards under natural conditions and that physiological constraints can limit the behaviours available to a species. These results should be compared with the results obtained by Taigen *et al.* (1982) on the aerobic dependence index in various species of amphibia (section 3.3.3). Both papers clearly show how physiological constraints can influence the distribution and abundance of animals and are good examples of how well-planned physiological ecology can contribute effectively to ecology and evolution.

Reptiles are well suited to studies on FMRs using DLW. The following case study is one of the most comprehensive of its kind and involved the measurement of seasonal changes in water balance, energetics, food consumption, daily behaviour, diet, osmoregulation and body mass of a desert tortoise, *Gopherus agassizii*, by Nagy and Medica (1986). The study was carried out in Rock Valley at the Nevada Test Site in Nye County, Nevada, and involved the use of DLW. The tortoises hibernated beneath the surface during winter and emerged in spring for about 3 hours every fourth day. During this period they had access to rainwater and succulent plant material and consequently gained weight while storing water in the urinary bladder. Surprisingly, however, energy intake, due to the high water content of the herbage, was not sufficient and total body solids actually declined. Without the use of isotopes this observation would not have been possible. The food plants eventually lost sufficient water in late spring and the tortoises were able to consume enough dry matter to ensure a positive energy balance. Body mass began now to decline because of the negative water balances the tortoises were experiencing (Fig. 3.25). This in turn caused an increase in the

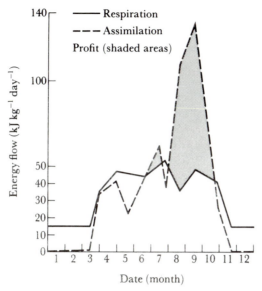

Figure 3.25   Seasonal energy budget of Rock Valley desert tortoises. Modified from Nagy and Medica (1986).

osmotic concentration of the bladder urine and blood plasma, and the urine became iso-osmotic with the plasma. As the environment became progressively drier during the summer, the tortoises ate less and spent more and more time aestivating in their burrows. Thunder showers in midsummer (July) allowed the tortoises to abandon aestivation and restore their water balance by drinking rainwater and voiding concentrated urine. These showers also allowed them to accumulate surplus energy as the grasses and forbs were still in a dry condition. The tortoises soon experienced osmotic stress, however, and once again exhibited a negative water balance and body mass declined until more rain in September provided relief. The animals again voided the concentrated urine, and stored dilute urine after copious drinking bouts. Energy balance during this period remained positive until late September (Fig. 3.25) when the high water content of the food plants once again prevented the tortoises from consuming sufficient energy, as occurred during spring. This situation continued until mid-November when winter hibernation started. The authors conclude that these animals are able to exploit ephemeral resources by relinquishing the maintenance of internal homeostasis on a daily basis for most of the year, but balancing their water and salt balances on an annual basis, while producing an energetic profit. Not only is this study by Nagy and Medica significant in itself but it provides a salutary lesson on the false impressions that the measurement of body mass alone can create, as well as the great potential of an integrated experimental approach which includes the use of DLW.

3.3.5 Birds

Birds are particularly well suited to nutritional and energetic studies. Many species can be tamed and kept in captivity with minimal stress. In fact the commercial chicken is probably the best-studied animal in the world from a nutritional point of view, because of its economic importance and its suitability as an experimental animal. It is disappointing, however, that zoologists seldom consult or study the excellent poultry science literature, whereas the poultry scientists work in isolation from avian ecologists and physiological zoologists. In addition to the relative ease with which birds can be captured, many species can be kept under almost constant observation which makes them suitable subjects for time-budget studies, and the fidelity with which they return to their nests makes them ideal subjects for energetic studies using DLW. The following case studies are a small sample of the very extensive literature that has accumulated on avian energetics and nutrition. Nevertheless, they should give the reader a general understanding and insight into the type of studies carried out in this field.

Walsberg (1983b) has published a comprehensive review on avian ecological energetics. His discussion centres on the most prominent activities and events in the avian life cycle, namely resource defence, gonadal growth and gametogenesis, incubation, care of the young and the moult. He also draws attention to the special characteristics of birds that influence their ecological energetics. For example, the high body temperatures and metabolic rates of most birds is not associated with the ability to store large amounts of energy. In fact, the opposite is true and it has been shown that a 25 g bird can theoretically store only enough energy for a maximum of 2 days of normal activity. Some birds can, however, store energy outside their bodies such as Clark's nutcracker (*Nucifraga columbiana*), which stores an amount of seeds in the autumn that is equivalent in energy value to 2.2–3.3 times the animals' energy requirements during autumn and winter (Vander Wall and Balda, 1977). Another prominent characteristic of birds described by Walsberg (1983b) is that of so-called determinant growth. In other words, most species attain their adult size relatively early in life and usually before becoming independent of their parents. Birds therefore channel their energy into fewer pathways than many ectothermic animals, which may be capable of storing large amounts of energy and exhibiting somatic growth throughout their lives. The annual cycle in birds is also relatively independent from previous and subsequent cycles, which facilitates the energetic analyses of the life styles of birds. Walsberg's (1983b) review contains many valuable summaries of the energetic cost of various activities in birds and the interested reader is referred to the original review for a more thorough analysis. Two figures from the review illustrating the relationship between body mass, basal metabolic rate and total daily energy expenditure as well as the relationship between body mass and the percentage of the active period spent in flight are

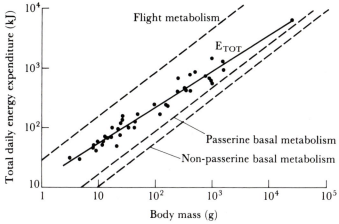

Figure 3.26   Relationship between basal metabolism, flight metabolism and total daily energy expenditure (E_{tot}) as a function of body mass. After Walsberg (1983b).

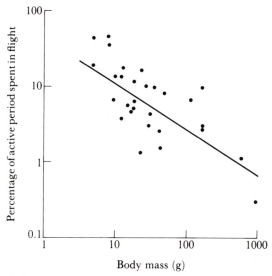

Figure 3.27   Relationship between body mass and the percentage of the active period spent in flight. Redrawn from Walsberg (1983b).

included in Figs 3.26 and 3.27 for interest. Walsberg concludes his review with the remarks that approximately half of a bird's total energy requirements are channelled to maintenance of body temperature and the basal metabolic rate. Because of the difficulty in measuring the thermal relationships of birds with their micro-environment under natural conditions, Walsberg is of the opinion that, with notable exceptions (Mugaas and King,

1981), the data in this field are of questionable accuracy and require further careful analysis before energy budgets can be accepted with confidence. Walsberg (1983b) also concludes that behavioural ecology will play an increasingly important role in energetic studies on birds and cites the example of weight loss during incubation in red junglefowl, studied by Sherry *et al.* (1980). These authors manipulated the food available to these birds experimentally during incubation and concluded that the weight loss which accompanies incubation is not a result of partial starvation, but rather a change in the set point around which body mass is regulated. This reduction in food requirements in turn allows the incubating female to reduce her time spent in foraging, and increase the time allocated to nest attendance. An interesting example but one could also describe the study as a physiological one as it involves nutrition and change in the tissue mass of the birds. Perhaps the strict demarcation between physiology and behaviour is a bit meaningless and therefore not particularly rewarding.

The study by Mugaas and King (1981), referred to above, consisted of a time-budget investigation of the annual variation in daily energy expenditure by the black-billed magpie. The investigation included a very careful and complex evaluation of the bird's heat exchange with the environment, which increased the overall accuracy of the study significantly. In this respect special emphasis was placed on the equivalent black-body temperature (T_e) and the flux of heat at the skin surface (H_m). Figure 3.28 shows which variables were used in calculating T_e and H_m as well as the thermal micro-environments available to the magpies. Some of the more important conclusions drawn by these authors were as follows. Postural changes alone altered the value of T_e under some conditions by as much as 11 °C within a single thermal micro-environment. The winter roost that was selected by the birds was within a micro-environment in which minimal convective and radiative heat loss would occur. The magpies could always avoid heat stress in the summer by sitting in the shade and the authors concluded that selection pressure in this species has favoured adaptation to cold rather than to heat. The energetic cost of thermoregulation was in any event low, about 5% of total daily energy expenditure (H_{TD}) when required. The cost of moulting was estimated at 8% of H_{TD}. The cost of activity, however, was responsible for most of the fluctuation in total daily energy expenditure during the annual cycle and varied from 25 to 50% of H_{TD}. The authors also noted that small differences in the time devoted by the birds to aerial activity resulted in large differences in the cost of activity per hour and therefore eventually in H_{TD}. Hence the observation by the authors that magpies tended to reduce energy expenditure to a minimum by restricting movement. These conclusions led to their hypothesis that, because behaviour is the greatest source of variation in H_{TD}, natural selection should favour behavioural patterns that allow a maximum return on the expenditure of time and energy for a particular activity. In this respect they found that foraging efficiency was

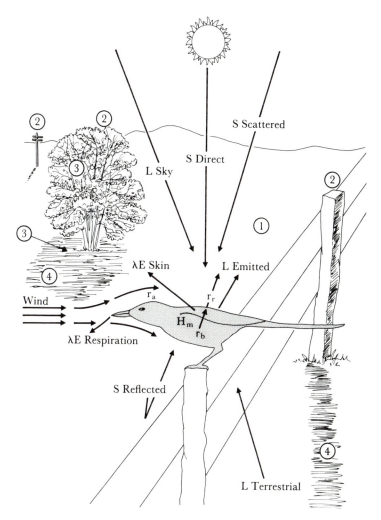

Figure 3.28 Variables used in calculating equivalent black body temperature (°C) and flux density of metabolic heat at the skin surface (W m⁻²) as well as the thermal micro-environments (1, 2, 3 and 4) available to the magpies. L, long-wave radiation; S, short-wave radiation; r_a, boundary layer resistance; r_b, whole body thermal resistance; r_r, radiative resistance; E, heat of vaporisation. Redrawn from Mugaas and King (1981).

high when food items were large, when the time spent in finding and swallowing food items was short and when energy intake was high (Mugaas and King, 1981). These results tempt one to speculate on the overwhelming importance of foraging efficiency for long-term fitness, but we should be cautious about accepting single activities or phenomena as being all impor-

tant in the complex interaction between the total life style of a bird and its
environment. For example, in a later study Paladino and King (1984)
showed that white-crowned sparrows used more energy sitting still in the
cold than while locomoting (see Chapter 1). Admittedly the latter results
were obtained for a small bird and endotherms with a greater mass would
not be as severely affected by cold. Nevertheless, while recognising the
obvious importance of optimal foraging patterns, caution is still indicated in
evaluating this variable in the overall energy budget and evolution of a
particular species.

In view of the wide acceptance of the DLW method as a reliable technique
to estimate the energetic cost of free existence, we will further examine the
remarkable versatility of this technique. The following examples of two
studies on very different species, the wandering albatross and the pied
kingfisher, should suffice.

Adams *et al.* (1986) estimated the energy expenditure of free-ranging,
wandering albatrosses (*Diomeda exulans*) on Marion Island, one of the two
sub-Antarctic islands of the Prince Edward Group. Wandering albatrosses
cover vast distances while foraging (as much as 2250 km) and, while feeding
chicks, these foraging flights can last on average for 6 days. Breeding males
and females spend 77% and 85% respectively of their total time budget
engaged in long foraging flights at sea. It follows then that any reduction in
the energetic cost of flight per unit time will benefit their energy budget very
significantly. This, in fact, seems to be the case as Adams *et al.* (1986) using
the DLW method established that the energetic cost of flight was only 2.35
times the measured RMR. This low value can be attributed to the relatively
low cost of the albatrosses' soaring flight (also known as oceanic soaring
along waves) when moderate to strong winds are blowing, a common
meteorological feature in these latitudes. Consequently, a bird's overall
energy expenditure of 3.354 kJ day^{-1}, measured by these authors, is only
1.83 times the RMR. This is reported as being the lowest FMR for a free-
ranging bird. Adams *et al.* (1986) have extrapolated their data to estimate the
energy expenditure of the total population of breeding, wandering albatrosses at the Prince Edward Islands (see Table 3.11). They have also
measured the energy content of the main food items of the albatross and, on
this basis, concluded that the total amount of food required by wandering
albatross adults and chicks at these islands is about 1.69 × 10^6 kg, comprising about 1.35 × 10^6 kg of squid, 0.17 × 10^6 kg of fish, and 0.17 × 10^6 kg of
crustaceans and carrion per annum. This excellent study will no doubt assist
nature conservationists and senior administrators greatly in making
decisions about future environmental protection in the Antarctic.

In sharp contrast to the very low FMR of albatrosses (1.83 × RMR), the
Cape gannet (*Morus capensis*) exhibits an FMR which is 4.7 × RMR. This is
probably because these seabirds use mainly flapping flight and indulge in
power dives when hunting their prey (Adams *et al.*, 1991).

(a)

(b)

Plate 3.3 The cost of free existence or field metabolic rate of the wandering albatross (a), *Diomeda exulans*, is far lower than that of the Cape gannet (b), *Morus capensis*. These large differences can be attributed to their very different life styles (see text for details); (a) shows how resting metabolic rate of a nesting albatross can be measured in the field. (Photos: courtesy Chris Brown and Nigel Adams.)

Table 3.11. Energy expenditure of breeding wandering albatrosses at the Prince Edward Islands. From Adams et al. (1986)

Activity	Male (9.44 kg)			Female (7.36 kg)			Total population size (pairs)	Population energy expenditure (kJ × 10⁶)
	Energy expenditure (kJ day⁻¹)	Time (days)	Total energy expenditure (kJ × 10²)	Energy expenditure (kJ day⁻¹)	Time (days)	Total energy expenditure (kJ × 10²)		
Prelaying	2 345	26	609	1 828	2	4	2 604	159.6
Incubation	2 855	44	1 256	2 226	34	757	2 425	488.2
Foraging during incubation	3 731	34	1 269	2 909	44	1 209	2 425	600.8
Brooding	2 345	16	375	1 828	16	293	2 245	150.0
Foraging during brooding	3 731	16	597	2 909	16	466	2 245	238.6
Chick rearing	3 731	246	9 179	2 909	246	7 161	2 184	3 568.7
Total								
Breeding cycle	N/A	382	13 285	N/A	358	9 890	N/A	5 206.0
Annual	N/A	365	12 964	N/A	365	10 083	N/A	4 950.0

N/A, Not applicable.

Our second example, namely that of the pied kingfisher, has been thoroughly studied by Heinz-Ulrich Reyer of Zürich University. Reyer has examined the role of energy expenditure by pied kingfisher parents in determining the acceptance of helpers by these birds at their nests in Kenya. Some 150 bird species have been reported to accept helpers at the nest and the pied kingfisher recruits two types of helpers: so-called primary helpers that help their own parents and secondary helpers that help birds other than their parents. Reyer and Westerterp (1985), using DLW, measured the daily energy expenditure of adult birds that fed nestlings. Concurrently, they recorded the number, type and size of fish fed to the nestlings throughout the activity period. This allowed them to calculate the average energy intake per nestling per day. The intensity of the nestlings' hunger (begging sounds) was also monitored by placing a microphone at the entrance of the nest chamber and recordings were made for 10 s every 5 min. These observations were carried out on two different breeding colonies, one of which was 'energetically stressed'. Later the authors manipulated clutch size, thereby reversing the begging and energetic stress upon the parents. They first found that energetically stressed parents in a colony with access to only a poor food supply accepted helpers more frequently than in the second colony where food was more easily available. Then, the experimental manipulation of clutch size and consequently energetic stress resulted in reversing these results. The authors also found that the duration of begging by the nestlings decreased significantly as their food supply increased and Reyer and Westerterp (1985) conclude that the demands of nestlings and the energetic stress upon parents are the most important proximate mechanisms determining the degree of helper recruitment. Previously, Reyer (1980) argued that this form of cooperative behaviour evolved through the combined effects of individual and kin selection and that the evolution was facilitated by a skewed sex ratio in favour of males and the fact that these birds breed in colonies. Also of interest is the finding of Reyer et al. (1986) that primary helpers have significantly lower blood titres of testosterone than secondary helpers. These low levels were also accompanied by smaller testes and aspermia. The authors ascribe this effect to behavioural domination by the breeding birds over the primary helpers: a form of psychological castration. They also advance the hypothesis that in the case of the primary helpers the cost of sexual competition with the breeding birds is higher than the benefits accruing from cooperative breeding. In the case of secondary helpers, however, who are not closely related and in whom testosterone levels are not reduced, the benefits of competition probably outweigh those of cooperation. The latter do not therefore show reduced fertility.

The next example is provided by the humming-bird, a bird which has attracted a great deal of scientific attention because of its small size and remarkable beauty. We have already noted in Chapter 1 that these birds exhibit nocturnal torpor because of their small size, large relative surface

(a)

area and associated high metabolic rate. In this example we shall first examine the energetic cost of hover-feeding, a prominent behavioural characteristic of these birds, and then learn how their digestive physiology determines their foraging frequency.

The energetic cost of hovering flight in humming-birds has been measured frequently since the pioneering studies of Pearson (1950) but probably the most innovative of these measurements is the recent study by Bartholomew and Lighton (1986). These authors measured the oxygen consumption of free-ranging Anna humming-birds (*Calypte anna*) in Los Angeles, California by modifying the feeding tube of a sugar-water feeder to act as a respirometry mask. Air was drawn at a constant rate into the mask and from there through CO_2 and H_2O scrubbers and a flow meter to eventually reach a sensitive oxygen analyser (resolution 0.001% O_2). The humming-bird had to insert its entire head into the transparent mask in order to feed and the presence of the bird's head within the mask was detected by a light-dependent resistor and recorded by a computer. The computer also sampled the oxygen content of

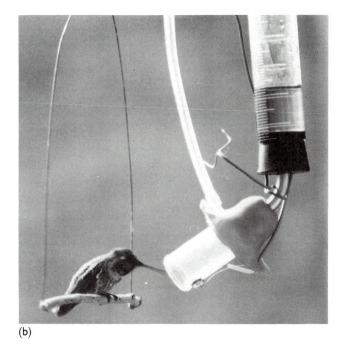

(b)

Plate 3.4 A novel technique for measuring the energetic cost of hovering flight in humming-birds under natural conditions. The nectar feeder has been modified to act as an oxygen mask (a) and the perch (b) is connected to an electronic device to record the mass of the bird. (Photo: courtesy John Lighton.)

the airstream continuously until it detected an oxygen depletion of 0.05%, whereupon it was programmed to retain the previous 140 samples in memory and then to record the next 500 samples. The mass of the birds was determined on a trapeze-like perch connected to a force transducer and suspended immediately in front of the feeder mask. This imaginative analytical and recording system allowed the authors to measure the mass of the animals, the length of their feeding bouts and the energetic cost of hover-feeding under natural conditions without any form of restraint or artificial stress upon the birds.

They found that feeding bouts consisted of numerous sallies during which the head was moved in and out of the feeding mask about once per second (see Fig. 3.29). The mean time spent by individual birds at the feeder ranged from 19.8 to 35.4 s while the number of sallies per feeding bout ranged from 9.5 to 27.4 s. The mean mass of the birds was 4.6 g and the mean energetic cost of hover-feeding was equivalent to 41.5 ± 6.3 ml O_2 g^{-1} h^{-1}. The latter value is very similar to values obtained by other authors for carpenter bees (2 g) and for different species of humming-birds ranging in body size from 3

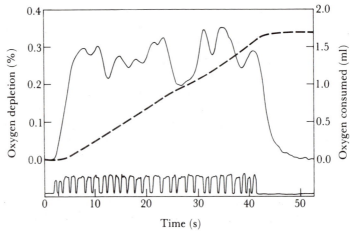

Figure 3.29 Percentage O_2 depletion (——), cumulative O_2 consumption (– – –) and photoelectronic trace of frequency of feeding sallies in a 4.4 g male humming-bird (bottom trace). Redrawn from Bartholomew and Lighton (1986).

to 10 g. Bartholomew and Lighton (1986) conclude therefore that their results provide support for the hypothesis that the power required for hovering flight is not dependent on body mass but rather on wing disc loading (see Epting, 1980 for fuller discussion).

The role of digestive physiology in determining the frequency of foraging bouts in humming-birds has been studied by Diamond *et al.* (1986). These authors used a new method for measuring the time taken for the crop to empty. This consisted of a double-isotope dilution technique in which birds were first fed 100 μl of ^{14}C-polyethylene glycol-labelled sugar-water via a capillary tube and thereafter fed an additional 10 or 100 μl of ^{3}H-polyethylene glycol-labelled sugar-water after 2–20 min. Five minutes was allowed for complete mixing of the two isotopes in the crop, before sampling the crop's contents by means of a 10 μl capillary tube. The dilution of the ^{14}C-polyethylene glycol in the sample reflects crop volume at time $t = 0$, and dilution of ^{3}H-polyethylene glycol measures crop volume at $t = 2$–20 min. These studies showed that a 100 μl meal passed through the crop within 15–20 min with a half-time of 4.1 min. The authors also measured the rate of absorption of D-glucose, L-proline and L-glucose *in vitro* by the humming-bird intestine.

They found that the rate of transport of L-proline and the Michaelis–Menten constant (K_m) for active glucose transport were similar to other vertebrates. However, the V_{max} for the active transport of glucose was higher than in any other vertebrate previously studied (1570 nmol min^{-1} cm^{-2}), whereas the passive permeability of the intestine to L-glucose was the lowest recorded for any vertebrate thus far.

In this way the humming-bird has evolved adaptations in its digestive physiology to cope with a high sugar diet. By increasing the density of glucose transport sites it enhances its ability to absorb large amounts of glucose and by reducing the passive permeability it may prevent loss of blood solutes to the fluid egesta and excreta in the lumen of the intestine (Diamond *et al.*, 1986). The authors used these results to explain why humming-birds feed 14–18 times per hour, each feeding bout lasting about 1 min, while spending most of their time resting on a perch. Diamond and his colleagues argue that the birds are using this time to digest their sugar-water diet as rapidly as possible and are not able to feed again until the crop has partially emptied from the previous feeding bout. The value established for the half-time emptying of a 100 μl meal from the crop, namely 4.1 min, means that a bird foraging with a half-filled crop would forage about 15 times per hour. This is in close agreement with the actual recorded rate of foraging of 14–18 times per hour. Diamond *et al.* (1986) conclude therefore that the humming-bird is maximising its energy intake within the constraints of its rate of digestion and that these birds are not minimisers of energy intake. The latter impression arose from the long periods of inactivity the birds exhibit between feeding bouts. Finally, the feeding ecology of humming-birds shows some interesting similarities to that of bumblebees that also spend long periods roosting in spite of an abundant nectar supply (see section 3.3.2).

Our final case study in this section also involves a fascinating bird–mammal interaction, namely that of honeyguides and human honey gatherers. The honeyguide is perhaps the only vertebrate animal capable of digesting wax and it supplements its insect diet by feeding on the larvae and wax from honey-bee colonies. The first assumption that the honeyguide was able to digest bees' wax arose from observations in East Africa, where these birds would fly into open-air churches to feed on the candles which in former times were made of pure bees' wax. Under natural conditions, however, the bird enlists the assistance of humans to break open the honeycombs while the gatherers keep the bees at bay with wood smoke. To achieve this the bird attracts the attention of the honey gatherer by flying in a restless pattern from perch to perch in close proximity to the person while emitting a characteristic call. It then leads the honey gatherer to the nest by means of a stereotyped behaviour pattern. Until recently most investigators were convinced that the honeyguide had no previous knowledge of the position of the bees' nests to which it was guiding the gatherers. Isack and Reyer (1989), however, have now shown clearly by means of carefully controlled field experiments in Kenya that the Boran honey gatherers are able to attract the attention of the honeyguides by means of a penetrating whistle and other noises. The Boran are then able to deduce the direction and distance to the nest from the bird's flight pattern. They are also able to tell when they have arrived close to the nest by observing another distinct change in the bird's behaviour pattern. These field experiments also showed that the birds had prior knowledge of the position and direction of the particular nest before commencing a guiding operation. How effective is guiding? Isack and Reyer (1989) conclude that guiding reduces search time by as much as 64%, thereby affecting the energy budgets of both gatherers and honeyguides very favourably. Without the help of the gatherers the birds would not easily gain access to the resource.

3.3.6 Mammals

Many case studies could be gleaned from the mammal literature but space only allows discussion of a few. These have been selected for their unusual rather than their representative nature.

The first case involves the study of nectar intake and energy expenditure in a species of flower-visiting bat, *Anoura caudifer*, which weighs only 11.5 g. This minute mammal has been studied by means of the DLW method in Venezuela by Helversen and Reyer (1984). These workers mist-netted the bats at the entrance to their roost, marked them and in the customary way injected them with $D_2{}^{18}O$ and collected the usual series of blood samples after release and recapture. They did, however, experience difficulty in withdrawing blood samples from such small and delicate creatures, which easily develop shock from robust handling. To overcome this problem they used a truly novel and innovative blood sampling procedure by placing the larva of a blood-sucking bug, *Rhodnius prolixus*, on the flight membrane of the bat. Within several seconds the larva located a suitable capillary with its sharp proboscis and within 4 min the bug larva was transformed into a plump ball filled with blood. It was then removed, decapitated and the stomach contents collected for analysis. It was not even necessary for the investigators to heparinise their collecting tubes as the larva had already secreted anticoagulant into its blood meal. The bats remained perfectly quiet and unperturbed throughout this procedure. Helversen and Reyer (1984) concluded from this study that these small flower-visiting bats exhibit a very high daily energy expenditure ($12.4 \, \text{kcal day}^{-1}$) and an even more dramatic daily water turnover rate of $13.4 \, \text{ml day}^{-1}$. The latter amount represents a daily water exchange equal to 115% of the body weight and 165% of the body water pool. These results remind one of similar problems faced by bumblebees and humming-birds, discussed previously in this chapter. They are also not unexpected in view of the rather dilute concentrations of nectar reported by these authors for neotropical bat flowers (range, 9–29%; $\bar{x}$, 17%). It would seem then that the dilute and dispersed nature of their diet, as well as their small size, has imposed a high energy turnover on these flying mammals and it is not surprising to learn that they fly with great speed and precision from flower to flower, in spite of being water-loaded.

From very small flying bats to overwintering bears. How do American black bears survive their winter retreat of about 5 months without food or water? During this retreat in its den, the bear does not enter true hibernation, as the body temperature falls only a few degrees. The females give birth and suckle their cubs during this period and the energy requirements of the bears are supplied by oxidation of large lipid reserves. The resulting metabolic water production compensates for the evaporative water loss and the bears remain in water balance throughout the winter, without the necessity for urinating. The only major nutritional and metabolic problems remaining are how to get rid of urea, the end-product of nitrogen catabolism in mammals, and to survive so long a retreat without any form of protein nutrition. Ahlquist *et al.* (1984) have investigated these problems by injecting labelled glycerol, one of the major end-products of lipid metabolism, and labelled urea into active and overwintering bears. The injection of labelled glycerol resulted in labelled amino acids, proteins, lactose and glucose appearing in the bloodstream, thereby confirming that glycerol was involved in the synthesis of amino acids and proteins. Radioactive urea appeared in active bears but not in the overwintering bears. The latter result was explained by following the fate of a labelled amino acid, alanine. The resulting radioactive urea from the catabolism of this amino acid appeared in active bears but not in overwintering bears. Ahlquist *et al.* (1984) conclude from these results that nitrogen from protein or amino acid catabolism in overwintering bears is diverted away from the ornithine cycle, which is responsible for the synthesis of urea, towards the metabolic pathways involved in amino acid synthesis. In this way the accumulation of urea is avoided and simultaneously new amino acids are synthesised. Urea recycling in ruminant animals serves a similar purpose, but involves the symbiotic micro-organisms in the rumen.

Our next case study is provided by mole rats belonging to the family Bathyergidae. Although these animals show some similarities to true moles, they are rodents and unlike the true moles, which are insectivores, these animals are vegetarian. This case study could also be entitled, 'The advantages of sociality in compensating for the wide dispersion of food resources under arid conditions', because those species that occur in arid habitats exhibit a surprising degree of eusocial behaviour. The ultimate example of social cooperation occurs in the naked mole rat which, like the social insects, is truly eusocial, with reproduction being suppressed in all females within the colony apart from in the large dominant 'queen'. This remarkable discovery of a eusocial mammal was made by Jennifer Jarvis at the University of Cape Town (1978). She has subsequently examined many aspects of the social life and physiology of these most unusual mammals but we shall limit our discussion to her first idea that colonial or social life in these animals was imposed upon them because of the limited food resources in arid areas and the high cost of tunnelling beneath the ground to obtain these resources. This original idea has been further advanced by Lovegrove in a series of significant papers (e.g. 1986) which culminated in the construction of a model which analyses the risks of unproductive foraging as a function of the dispersion of resources and the group size of the foragers (Lovegrove and Wissel,

Plate 3.5   The naked mole rat (*Heterocephalus glaber*) lives in warm underground tunnels in the dry parts of Kenya. It is a poikilothermic mammal and has evolved a eusocial life style to make best use of the widely scattered food resources in these arid areas. (Photo: courtesy Jennifer Jarvis.)

1988). Briefly, Lovegrove maintains that as the habitats of the various species of mole rats in southern Africa become more arid, the species become progressively smaller and more social in their life style. He explains this trend on the basis of the distribution and size of the underground tubers or geophytes upon which these animals feed. Figure 3.30 shows that as aridity increases from south to north in southern Africa these geophytes become larger and more widely dispersed. If one then accepts Vleck's (1979) estimate that the energetic cost of burrowing beneath the ground can be 3600 times the cost of walking on the surface, then obviously mole rats in arid areas run great risks when expending large amounts of energy in burrowing because of the low probability of a successful encounter with a large tuber. Clearly then it would be of great advantage to the arid-dwelling species to have more foragers per colony to increase the probability of encountering a tuber. Equally obvious, once the tuber is encountered it must be shared—hence one of the reasons for a social life style (Fig. 3.31). Moreover, it would not be to the advantage of the colony to merely increase the number of foragers as this would increase the total demand for energy in an arid area with limited resources. An alternative would be for natural selection to produce more foragers with a much smaller body size.

The evolution of a larger number of foragers with small body size reduces the foraging risk geometrically (see Lovegrove and Wissel's model in Fig.

Increasing aridity

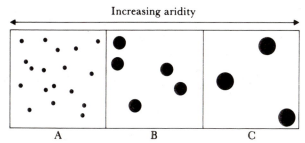

A   B   C

Figure 3.30 Diagrammatical representation of the general trend, with increased
aridity, of increased plant tuber size and increased nearest-neighbour
distance. Tuber distributions are markedly random for the situation
depicted in (B) and (C), but are unknown for the mesic situation
depicted in (A). Redrawn from Lovegrove and Wissel (1988).

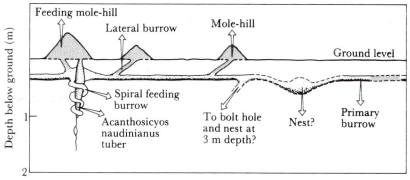

Figure 3.31 A schematic diagram of the burrow types and burrow structures
comprising the burrow system of *Cryptomys damarensis* at a dune
site. Broken lines indicate parts of the system that were not exca-
vated. Vertical axis to scale only. Redrawn from Lovegrove and Painting (1987).

3.32). Nevertheless, a reduction in body size also has inherent disadvantages
from an energetic point of view, namely the predictable increase in mass-
specific metabolic rate which occurs with decreasing body mass. However,
much to Lovegrove's delight (1986) the resting metabolic rate of these small,
social and arid-adapted mole rats is far less (43%) than the predicted value
and Lovegrove and Wissel (1988) report the surprising fact that the mass-
specific resting metabolic rate in the Bathyergidae does not scale in the
normal way but tends towards mass-independent scaling (Fig. 3.33.). This
represents a significant saving in energy while still allowing the many advantages of increasing the number of foragers with smaller body size. Lovegrove
and Wissel have termed this metabolic adaptation 'risk sensitive metabolism'

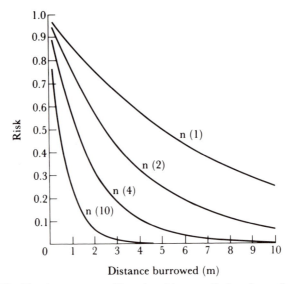

Figure 3.32 The decay curves of foraging risk, quantified as the probability of no encounter with a plant tuber, as a function of distance burrowed for different numbers (*n* of cooperatively foraging mole rats). The data show rapid reduction in risk with increasing numbers of cooperatively foraging mole rats. After Lovegrove and Wissel (1988).

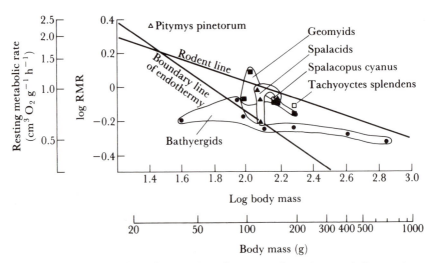

Figure 3.33 Mass-independence scaling of mass-specific resting metabolic rates in bathyergid mole rats compared to standard rodent curve. For original references and data see Lovegrove and Wissel (1988).

as it is a basic adaptation in ameliorating the high risks of foraging in the arid habitats of mole rats.

3.4 Concluding remarks

If anyone wished to be convinced about the complexities of the natural associations among organisms on this planet, they need only contemplate the relationships between the myriad animal species and their food plants, from whence they obtain their energy. These relationships, through natural selection, must have changed fundamentally over millions of years before evolving into the food chains and webs which we observe around us today. Plants have evolved special attractants, such as fragrances, nectar and pollen to attract pollinators, thereby ensuring greater genetic variation within a species. Dispersion of seeds has been enhanced by encasing seeds in nutritious arils or the brightly coloured flesh of sweet fruits. In contrast, plants have also developed complex defence systems including toxic chemicals to deter herbivores from destroying both the vegetative portion and the seeds of plants. Before seeds ripen the unripe fruits usually contain chemical deterrents to prevent premature dispersal. This chapter has commented on several examples of these defence mechanisms as well as the consequent adaptations evolved by the herbivores to overcome these defences in an 'arms race' between plants and herbivores. Nutrition and therefore the nutritional niches of animals are naturally of central importance in their evolution and ecology and we can expect studies of the above phenomena to become preeminent in the near future.

Studies of food chains and webs tend to be mostly of a descriptive nature, and allow ecologists to trace how energy flows through ecosystems. When examining these food chains it is always tempting to generalise on the evolutionary significance of how and why a specific kind of animal fits into a certain trophic level or node within a web. Attempts of this nature will reveal for example that, as one proceeds up the food chain, animals usually increase in size by large increments from one trophic level to the next. In other words, a saltatory rather than a uniformly smooth increase. There are, however, many interesting exceptions, e.g. the wolf is smaller than its prey whereas the baleen whale, feeding on microscopic plankton, should in theory occupy a much higher position in the food chain. Evolutionary explanations of the flow of energy in food chains and webs, particularly with respect to the partitioning of resources and competition among species, should therefore be approached with great caution in view of the very complex nature of the phenomena involved, and the great difficulty of manipulating natural ecosystems to answer even simple questions about interspecific competition.

It is also tempting to generalise about how the use of energy has evolved in very different ways among various groups of animals. For example, certain animals, such as bumblebees, honey-bees, diminutive bats and humming-birds are extravagant in their use of energy. In fact, so much so, that even the water they consume in their sugar diet has become a constraining factor in limiting their activity. This energetically expensive life style naturally allows them to obtain rich rewards in the form of nectar and pollen and to fill specific niches within the nutritional mosaic. In contrast, the polar bear lays down huge reserves of fat and uses them sparingly by employing long periods of semi-torpor. Although the above differences can largely be explained on the basis of great differences in body size, the large African antelopes, in spite of their size, have very limited fat reserves. Their insulation against heat loss is minimal and cold, wet conditions after long periods of drought often result in extensive mortality. Body size is therefore not an entirely acceptable explanation and, as we have seen from Chapter 1, natural selection for facilitating heat loss in African antelopes appears to have been a dominant factor in their evolutionary history, forcing them to walk an energetic tight-rope in many instances. Nevertheless, it is possible to some extent to general-ise about the availability of oxygen in the environment (see Fig. 3.9) and how this is associated with the distribution of metabolic patterns and the life styles of animals. In this regard we need only compare the metabolic rate of a sluggish mud-dwelling arthropod with that of say a bumblebee. Graham (1990) has examined the evolutionary and ecological implications of these associations.

The nutritional value of plants varies but an essential and important component of all plant food is cellulose, a polymer of glucose. It is therefore most surprising that only a few animals have evolved the specific endogenous enzyme, cellulase, to break cellulose down to its constituent sugar molecules for use as energy. Instead, we find that complex digestive systems such as ruminant and caecal digestion have evolved to allow microbial digestion of cellulose. This may seem to be an unexpected evolutionary direction, but when we evaluate the many advantages of microbial fermentation, such as vitamin synthesis and protein synthesis from non-protein nitrogen (urea), perhaps an understandable development.

Most of the examples discussed in this chapter have involved aerobic respiration. The importance of anaerobic respiration should, however, not be underestimated and the examples given of how the life styles of different species of frogs can be explained to a large extent on their relative use of anaerobic vs. aerobic respiration are very significant. These case studies are also among the best examples of good correlations between physiological attributes and the fitness of particular species. The evolution of endothermy and the concomitant quantum jump in aerobic capacity, however, allowed the development of sustained high activity with greatly improved endurance. Animals that evolved these physiological attributes could occupy entirely new niches and, in some cases, came to dominate ectothermic species. This does not mean that endothermy and high aerobic scope are necessarily a more efficient form of energy use. In fact, the very opposite is true and studies on the FMR of certain lizards show that these animals require 17 times less energy than an equivalent sized bird or mammal for free existence. Even the net energetic cost of limbless, lateral locomotion by the snake *Coluber constrictor* is the same as the net energetic cost of running in birds and mammals (Walton *et al.*, 1990). Both the ectothermic and endothermic life styles have their respective advantages and disadvantages and the environmental constraints imposed by the niche will determine which metabolic pattern will be most appropriate. For example, it is not surprising that in the desert with abundant sunshine and limited energy resources the small ectotherms (arthropods and reptiles) are the most abundant animal groups.

Scaling of metabolic rate, since the earliest work of Kleiber on his mouse to elephant curve, has provided both ecologists and evolutionists with endless speculation. Many questions in this regard remain unanswered but the general concept has been most useful in both theoretical and practical applications. The recent findings of Lovegrove and Wissel (1988) that fossorial mole rats, exposed to arid environments where food tubers are patchily distributed, tend to decrease in size and become eusocial are very significant in this regard. The large number of small social foragers provides a much higher probability of a forager encountering a tuber, but the smaller size has not resulted in the expected increase in mass-specific metabolic rate. In these animals, then, the resting metabolic rate tends towards mass-independent scaling, which has obvious advantages for the survival of this species in its energetically demanding niche. Similarly, the golden mole of the Namib dunes, *Eremitalpa granti namibensis*, swims through the soft dune sand in search of its prey of mostly termites and insect larvae. This exceedingly costly form of locomotion has resulted in these animals exhibiting high thermal conductance, a basal metabolic rate much lower than the predicted value and the equivalent of daily torpor (Fielden *et al.*, 1990). In contrast, the cost of flight in certain insects and birds is so high that predictions based on body mass become difficult. However, it is of interest to note that the maximum size that a flying animal can theoretically reach depends largely on gravity and air density. The fossil evidence of giant flying animals suggests, together with other evidence, that gravity may have varied widely over geological time.

These concluding remarks would be incomplete without emphasising the very significant improvement that has occurred in studying the energy exchange of animals as a result of the development of the DLW technique. This allows the concurrent measurement of FMR and water turnover rates under natural, free-ranging conditions and its advantages have been well illustrated in the excellent study of the ecological energetics of desert tortoises by Nagy and Medica (1986), described in section 3.3.4. The far more realistic values obtained by studying FMRs as opposed to laboratory studies of gas exchange have already revised our traditional views on the physiological ecology of several animal species. It will continue to do so.

The ecological energetics of humans have received considerable attention both from a medical, sports and industrial viewpoint. It appears that our energy requirements are reasonably close to the values predicted by our body mass, apart from our rather inefficient bipedal locomotion. However, that is where any similarity with other mammals ends. Man is the only species to have discovered the secrets of lever mechanics, which have subsequently led to the discovery of how to transform fossil fuels into various forms of mechanical and electrical energy. It is a fascinating exercise to trace how man's usage of energy has influenced the course of history from the building of the pyramids and cathedrals, through modern agriculture, the construction of vast railway systems and nuclear weapons to ultimately the Apollo rockets that placed the first humans on the moon. This reckless use of energy has had profound effects on the ecology of our planet, such as the greenhouse effect, to name but one.

One of the reasons why man has been so successful in exploiting the energy reserves in the environment is due to a social life style which includes the division of labour. Similarly, several of the case studies in this chapter have shown how important sociality is among animals to overcome situations where energy is limiting or periodically scarce. We noted in Chapter 2 that sociality was in many ways the ideal answer to water stress in arid environments. The same appears to be true in the case of nutritional or energy stress and here we are not only referring to the social insects but also to kingfishers, mole rats and even desert beetles belonging to the family Tenebrionidae (Rasa, 1990). It is also significant to note that subsociality has only developed in those isopod species living under the demanding conditions of a desert habitat. The best-studied species in this regard is *Hemilepistus reaumuri* and its ecology, kin recognition and evolution have been well described by Linsenmair (1987). The social way of life is therefore far more widespread than was earlier believed and has, in modified form, even been observed among spiders belonging to the genus *Stegodyphus*, which inhabit dry thorn-brush habitat on the African savannah. Seibt and Wickler (1988) have studied two species belonging to this genus in detail and have found that they form colonies of up to several hundred individuals. They cooperate in nest and web building as well as catching prey but compete for food. The communal nest provides protection against unfavourable climatic factors and predators. The authors are of the opinion that sociality within this genus evolved via neoteny and not from the mother–young bond.

To conclude, there are myriad ways in which energy is used by various animal species and this fact, together with the innumerable functions, both behavioural and physiological, which require energy, preclude us from making sweeping generalisations about the evolution of metabolic patterns and the associated ecological energetics. To name but a single example, Koehn (1987) has shown that bivalves with genetically higher leucine aminopeptidase (LAP) activity adapt more rapidly and effectively to osmotic stress, but higher LAP activity causes a higher protein turnover when the animals are starved, which has meant that selection against this trait has occurred because of the overriding importance of nutrition. We are again reminded of the importance of first appreciating the total ecology of a species before evaluating an adaptive character. Nevertheless, there is probably no single physiological characteristic, other than energy metabolism, which is a more sensitive measure of an animal's life style and particularly its function within an ecosystem.

CHAPTER 4

REPRODUCTION AND THE
ENVIRONMENT

The 'fitness' of any individual, population or species in the evolutionary sense is ultimately determined by its reproductive success. Reproductive rate need not necessarily be very high or at very frequent intervals but it must be successful. Consequently, it is not surprising that many of the countless reproductive patterns of animals are strongly influenced by environmental factors to ensure the most favourable environmental conditions for the completion of the reproductive cycle of a species. Before we examine these factors and how they interact with the reproductive physiology of individual animals, it is first necessary to gain an understanding of the delicate and rather complex endocrine control systems governing the reproductive patterns of animals. However, in view of the tremendous variation among animals in this regard, space will only allow a discussion of three examples, which have been chosen to illustrate both differences and similarities. In the case of the mammal and insect examples they also represent well-studied animals about which a considerable knowledge has accumulated.

4.1 Endocrine control of reproduction

4.1.1 Mammal

The endocrine control of reproduction in mammals has been well studied because of the generous funding available for medical and agricultural research. Sheep have proved to be particularly suitable experimental animals for this purpose and the following account of the sexual cycle in the ewe should provide the reader with a good understanding of the physiology of mammalian reproduction. We should, however, perhaps first refresh our memories of the anatomy of the female reproductive system and the associated endocrine structures (see Fig. 4.1).

The endocrine control centre is situated in the brain (hypothalamus), or more precisely that portion of the hypothalamus known as the median eminence. The median eminence is situated on the floor or ventral portion of the brain and lies just above the all-important pituitary gland. Specialised cells in the median eminence synthesise simple peptides known as releasing hor-

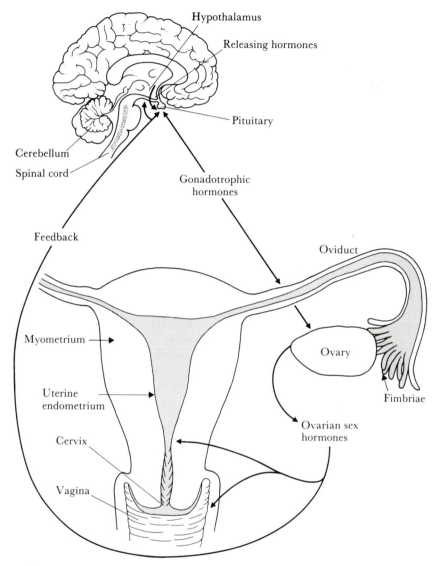

Figure 4.1 Major anatomical structures involved in the sexual cycle in a mammal.

mones and these are transported from the median eminence to the pituitary gland via an intimate closed system of blood vessels, known as the hypophyseal portal blood system. When the releasing hormones reach their specialised target cells in the anterior pituitary gland, they cause the release of vitally important pituitary hormones in the correct sequence, so that the reproductive cycle can proceed through its various stages. These so-called gonadotrophic hormones from the pituitary gland regulate the secretion of sex hormones from the ovaries in such a way that the reproductive tract of the female (uterus and vagina) is brought into the correct condition for implantation of the fertilised zygote, as well as being responsible for many other important changes. The delicate interplay of releasing hormones from the brain, the gonadotrophic pituitary hormones and the ovarian sex hormones control the reproductive cycle from mating to lactation. The following discussion explains how this interplay takes place in the ewe.

Ewes that belong to temperate sheep breeds are seasonal breeders and lamb during spring, which is usually the most favourable time of the year. They must therefore become sexually active and mate in the autumn to produce lambs in the spring. The environmental stimulus responsible for stimulating sexual activity in the autumn is the declining photoperiod or daylight length. This change in the photoperiod is perceived by the optic nerves and then a still unknown mechanism stimulates the pulsed release of a simple peptide from the hypothalamus, known as gonadotrophin releasing hormone (GnRH). This hormone is released by that portion of the hypothalamus known as the median eminence.

The GnRH is transported down the hypophyseal portal blood system to the anterior pituitary where it causes the release of the gonadotrophic hormone known as follicle stimulating hormone (FSH), as well as a little luteinising hormone (LH). The FSH enters the general circulation and arrives in the ovarian tissue where it results in the recruitment of follicles and the stimulation of their growth. When the follicles reach medium size the thecal cells surrounding the basement membrane of the follicle start to produce male hormones (androgens) and to develop receptor sites for LH. The so-called granulosa cells at this stage are not able to produce any steroid hormones (see Fig. 4.2).

The next step is further maturation of granulosa cells by FSH until they are able to transform the androgens produced by the thecal cells to female sex hormones, known collectively as oestrogens. The chemical reaction involved is known as the aromatising reaction. The oestrogens thus formed are powerful mitogens and cause rapid proliferation of granulosa cells. The follicle now grows rapidly and increasing levels of oestrogens are secreted. Finally, through the combined action of FSH and oestrogens, the granulosa cells develop receptor sites for LH on their cell surfaces.

The increased amount of oestrogens now enters the general circulation and has many profound effects on the physiology of the ewe: thickening of the vaginal wall and increased mucus secretion from this tissue; primary growth of the inner lining (endometrium) of the uterus is stimulated; the smooth muscle wall of the uterus (myometrium) becomes more 'irritable', i.e. its threshold for contraction is lowered; and finally and very importantly the oestrogens induce 'behavioural oestrus', commonly known as the 'heat' period. This is achieved by the action of oestrogens on the central nervous system and is the period when the female is receptive to the male and will

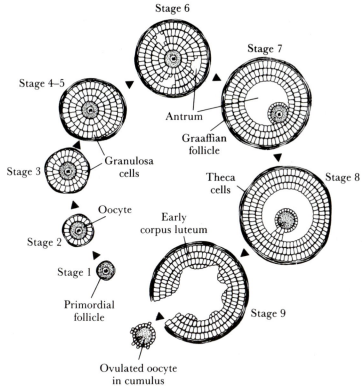

Figure 4.2 Eight stages of follicular growth in a mammal. Redrawn from Austin and Short (1972).

allow mating to take place. The oestrus period lasts for about 24 hours in the ewe and ovulation occurs towards the end of this period.

The high levels of circulating oestrogens also reach the anterior pituitary gland and change the so-called 'steroid environment' of this gland. This has important consequences because GnRH, which up till now has stimulated mostly FSH, now causes a sudden surge of LH release. This surge causes ovulation, luteinisation of the granulosa cells and formation of the corpus luteum, which is now capable of producing large amounts of progesterone. Prostaglandins are responsible for the actual rupture of the follicle and release of the ovum during ovulation (see Fig. 4.3).

During mating seminal fluid and spermatozoa are ejaculated into the vagina of the ewe and sperm arrive very soon in the oviduct, where fertilisation of the ovum takes place. The speed with which this happens cannot be due to the swimming speed of sperm alone, and it is thought that contractions of the muscles of the uterus (myometrium) may assist the passage of sperm towards the oviduct. It is also possible that the prostaglandin hor-

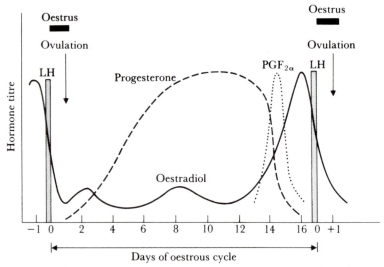

Figure 4.3   The sequence of hormonal secretions in the oestrous cycle of the ewe. Redrawn from Goldsworthy *et al.* (1981).

mones contained in the seminal fluid (the site of their first discovery) may be partially responsible for causing the contractions of the uterus, which has previously been primed by oestrogens. Another important action of oestrogens is to stimulate the swelling of the terminal portions of the oviducts, the fimbriae, which then clasp the ovary, thereby guiding the released ova into the oviduct. Nevertheless, trans-abdominal migration of the ova to the contralateral oviduct can occur in some species. Once the ovum arrives in the oviduct it is transported towards the uterus by cilia beating in that direction. The sperm swimming in the opposite direction are apparently too small to be affected by the beating cilia.

After ovulation takes place, the progesterone produced by the corpus luteum causes the final development of the endometrium, which now takes on a lace-like appearance when viewed under the microscope. The endometrium is now in an optimum condition for reception and implantation of the blastocyst. Irritability of the myometrium falls.

If fertilisation does not take place the uterine endometrium is stimulated by progesterone to produce prostaglandin $PGF_{2\alpha}$ which, via the uterine vein and the ovarian artery, reaches the corpus luteum and causes rapid luteolysis. Thus the corpus luteum programmes its own demise. The cycle then repeats itself with GnRH causing FSH release in the now 'progesterone-dominated' steroid environment of the pituitary (see Fig. 4.4).

However, if fertilisation does occur, the presence of the embryo within the uterus prevents lysis of the corpus luteum. The exact mechanism is still not known but either $PGF_{2\alpha}$ release is prevented or luteal tissue becomes more

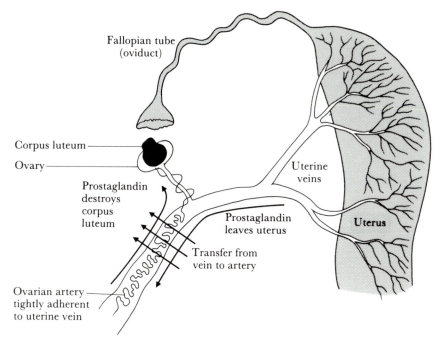

Figure 4.4 The vascular route whereby prostaglandin $F_{2\alpha}$ reaches the corpus luteum where it causes lysis of the corpus luteum. Redrawn from Austin and Short (1972).

resistant to it. In primates chorionic gonadotrophin is secreted from the placenta 6 days after conception and this may ensure survival of the corpus luteum. This is then the first example of the control of the maternal endocrine system by the embryo, known as 'foetal autonomy'.

Gestation or pregnancy then proceeds to term under the dominating influence of high blood levels of progesterone, which maintain a low level of irritability of the myometrium. In some species, for example the horse, the high levels of progesterone are maintained by the formation of additional corpora lutea of pregnancy. In others, large amounts of progesterone are produced by the placenta. The placenta in many species produces chorionic gonadotrophin that is mostly LH and is responsible for stimulating progesterone secretion by the corpus luteum. During pregnancy oestrogens are still secreted but in much smaller amounts than progesterone. One of the functions of these circulating oestrogens is to stimulate duct development in the mammary glands while progesterone stimulates the development of the secretory alveoli in these glands.

When full term is reached parturition must ensue and the control mechanism involved is again an interesting example of primary control by the foetus over the maternal endocrine system. Once the hypothalamus of the foetus

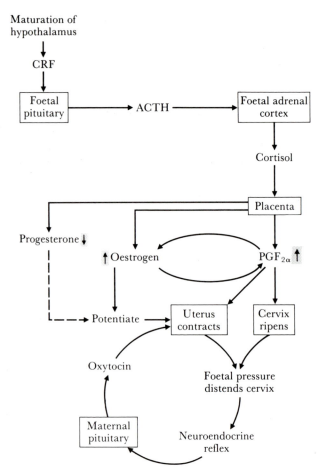

Figure 4.5   The sequence of hormonal events involved in parturition in the ewe.
See text for full details. Redrawn from Goldsworthy *et al.* (1981).

has matured it apparently secretes corticotrophin releasing hormone, which in turn causes the release of adrenocorticotrophin by the foetal pituitary. Adrenocorticotrophin then stimulates the synthesis and secretion of cortisol by the foetal adrenal cortex. The cortisol stimulates the synthesis of enzymes in the placenta which are capable of converting progesterone to oestrogens, and blood levels of progesterone fall while oestrogens rise. The oestrogens cause myometrial irritability and stimulate myometrial contraction. Foetal cortisol also stimulates placental synthesis and release of $PGF_{2\alpha}$ which has several profound effects:

1.   increases oestrogen biosynthesis by inducing aromatisation enzymes. Oestrogens in turn stimulate $PGF_{2\alpha}$ release (a positive feedback loop);

2. causes lysis of the corpus luteum and hence decline in progesterone synthesis and release;
3. stimulates uterine smooth muscle to contract;
4. ripens and dilates the cervix (see Fig. 4.5).

Once the above events have been entrained the mechanical distension of the uterus and cervix by the foetus causes, via the central nervous system, a release of oxytocin from the maternal neurohypophysis (posterior pituitary) and powerful contractions of the uterus and expulsion of the foetus takes place (third example of foetal control over the maternal endocrine system).

Note that in anencephalic foetuses with underdeveloped hypothalami and in pregnant ewes feeding on the alkaloid-containing skunk cabbage (which causes deformities of the hypothalamus) gestation is greatly prolonged, often with fatal consequences.

Shortly after parturition, circulating blood levels of both oestrogen and progesterone fall dramatically and prolactin release inhibiting hormone (PIH) in the hypothalamus is no longer secreted. Then, under the influence of the suckling stimulus, prolactin releasing hormone from the maternal hypothalamus is responsible for prolactin release from the anterior pituitary and milk synthesis ensues. (Another example of the progeny providing primary control over the maternal endocrine system.)

Finally, the suckling stimulus via the central nervous system causes release of oxytocin from the neurohypophysis, which stimulates the contraction of the myoepithelium in the mammary glands. In this way milk is ejected into the mouth of the suckling newborn lamb, known as the 'milk ejection reflex'. For this reason zoologists employ oxytocin injections to collect milk samples from wild mammals.

Male mammals in certain species also exhibit seasonal breeding and their periods of sexual activity are also controlled by a series of stimuli and feedback responses involving the central nervous system, the environment and the endocrine system. Not unexpectedly, the interplay of neurosecretions and hormones in the male is less complex than in the female, as is borne out by the following brief description.

Once the appropriate environmental trigger has been perceived by the central nervous system, the signal is integrated and GnRH is secreted by the hypothalamus and transported by way of the hypophyseal portal blood system to the anterior pituitary gland. Here GnRH causes the release of both LH and FSH from the anterior pituitary. These pituitary hormones are identical to those secreted by the female but naturally have different functions on the male gonads. In the male, LH stimulates the Leydig or interstitial cells to produce the steroid hormone testosterone. These interstitial cells are located between the coiled seminiferous tubules in which spermatogenesis takes place, and the testosterone they produce has a variety of well-known secondary effects in male vertebrates. These include, for example, stimulat-

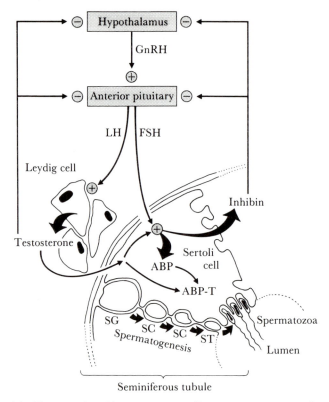

Figure 4.6   The interplay of hormones controlling spermatogenesis and androgen (testosterone) production in the male mammal. SG, spermatogonia; SC, spermatocytes; ST, spermatids; ABP, androgen binding protein. After Goldsworthy *et al.* (1981).

ing the development of the external genitalia, hardening of the antlers of deer and the development of the splendid male plumage in some birds and their accompanying aggressive male behaviour. Perhaps less well appreciated is the effect of testosterone on spermatogenesis. Once FSH has initiated the spermatogenic cycle, it appears as if testosterone plays an important role in maintaining spermatogenesis by its effect on Sertoli cells located within the lumen of the seminiferous tubule (see Fig. 4.6). In this regard FSH stimulates the synthesis of androgen-binding protein (ABP) in the Sertoli cells and once ABP has bound the testosterone it creates a high localised concentration of testosterone in the Sertoli cells, which is apparently essential for spermatogenesis.

Feedback systems are also in operation in the endocrine control of male reproduction. Testosterone blood levels, if too high, depress hypothalamic output of GnRH and apparently reduce the sensitivity of the anterior pituitary to GnRH. In the case of FSH, the evidence for feedback is not quite so convincing but it is thought that the Sertoli cells synthesise a peptide known as inhibin under the influence of FSH and that inhibin is then responsible for negative feedback on the anterior pituitary and perhaps on the hypothalamus (Fig. 4.6).

Once again we are presented with an excellent example of how effectively the hypothalamus integrates environmental stimuli to control important physiological responses. If we now add some of its many additional functions such as control of thermoregulation, appetite and osmoregulation, it is truly remarkable that a portion of the central nervous system, weighing only a minute percentage of the total body weight of a human, can accommodate so many vital control systems, including greed, lust and thirst!

It is also clear that the endocrine control of the reproductive cycles in mammals involves a large number of tissues, glands and hormones which are controlled by an exquisitely timed sequence of biochemical events involving stimulus, interaction and feedback. We now move to a discussion of the endocrine control of reproduction in an insect which, although very different, nevertheless involves certain similar mechanisms, particularly with respect to the control of ovarian function by the central nervous system.

4.1.2 Insect

There is no such thing as a typical insect. Some insects, such as the domestic cockroach and the blood-sucking bug *Rhodnius*, are easily kept in captivity and lend themselves to experimentation. Hence a considerable body of knowledge on their endocrinology has been documented. This is slightly less true for the mosquito, but because of its widespread distribution and ecological importance we have chosen it as our example for study.

Anyone who has camped outdoors or done field work beside the scenic lakes and forests of North America will vouch for the sudden and exasperating population explosion of mosquitoes that occurs in early spring. Throughout the winter the adults or fertile eggs, containing fully developed larvae, have remained frozen and dormant near the ice-cold waters of ponds, streams and lakes. The eggs must be able to tolerate freezing and are probably protected by cryoprotectants although very little is known about this aspect. As soon as temperatures rise and the oxygen concentration of the water falls below a threshold value, the eggs hatch within several days, the time depending upon the species involved. The emerging larvae feed on microscopic particles in the water and are air breathing. In some species (*Culex*) air is obtained through a breathing tube or siphon attached to the last abdominal segment, which is held just above the water surface. In others (anophelines), the larvae rest with the length of their bodies along the surface while exchanging gas directly through spiracles on the eighth abdominal segment. After approximately 3 weeks the larvae will have moulted three

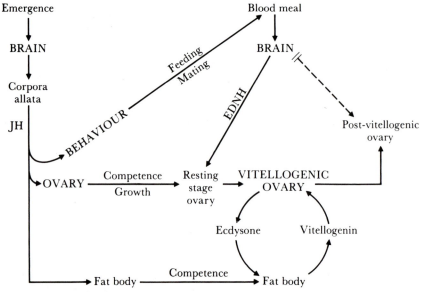

Figure 4.7   Environmental factors and hormones controlling egg development in the mosquito *Aedes aegypti*. JH, juvenile hormone; EDNH, ecdysiotrophic hormone. After Hagedorn (1983).

times and become sufficiently mature to enter the pupal stage. The moulting hormone ecdysone will have gained ascendancy over juvenile hormone to control this final development. The pupae do not feed but rest near the surface while final development of the adult is taking place. During this phase air is taken in through two minute 'ear-trumpets' which are situated on the thorax just posterior to the head. Eventually the pupal skin splits after several days at the interface between the air and the water and the winged adult escapes into the air. In this regard it is significant that the temperature at which the larvae develop affects the ultimate size of the adults; larvae at 20 °C produce significantly larger adults than those reared at 27 °C.

At the time of adult eclosion, primary follicles are already present in the mosquito ovary and they consist of an oocyte and seven nurse cells surrounded by follicle cells. The subsequent endocrine control of oocyte development leading ultimately to oviposition has been reviewed by Hagedorn (1983) and is summarised in Figs 4.7 and 4.8. Juvenile hormone from the corpora allata is the first hormone to initiate oogenesis and, during the first 2 days after eclosion, stimulates the primary follicle to double in size and to develop the ability to secrete the steroid hormone ecdysone in response to ecdysiotrophic hormone. To begin with, however, the follicles remain in the so-called resting stage until the first blood meal is taken. The presence of the blood meal in the stomach initiates the synthesis and release of the proteolytic enzyme trypsin for digestion of the blood meal, as well as the release of

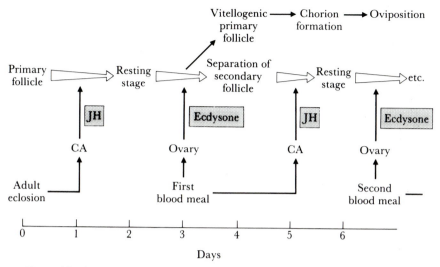

Figure 4.8 Sequence of physiological events involved in egg development cycles in the mosquito from eclosion to oviposition. After Hagedorn (1983).

ecdysiotrophic hormone from the brain. This in turn causes the production of ecdysone by the ovary. The resulting ecdysone then entrains vitellogenin synthesis by the fat body, which allows the development of yolked or vitellogenic follicles, leading to chorion formation, mating and ultimately oviposition. Mating patterns vary from species to species and naturally depend on specific mate recognition signals between the sexes. It is usually a very brief procedure lasting a few seconds. The male does not feed on blood and little is known about the endocrine control of the male sexual cycle. Oviposition, however, also requires further hormonal stimulus. Klowden and Blackmer (1987), for example, have shown that both gravid and non-gravid mosquitoes, injected with the haemolymph of gravid females, exhibit preoviposition behaviour by being attracted to 'oviposition-site' stimuli. In other words, they are in a physiological condition to start searching for a suitable site at which to lay their eggs. In contrast, ovariectomised, blood-fed females failed to respond to these stimuli. However, the re-implantation of as few as six mature follicles could induce pre-oviposition behaviour after a blood meal. In addition, the act of mating or, more specifically, the deposition of male seminal fluid in the female mosquito, is essential before preoviposition behaviour ensues under natural conditions.

It should also be noted that juvenile hormone is again secreted 2 days after the blood meal and the secondary oocyte now develops to the resting stage until the second blood meal is taken and the cycle repeats itself (Fig. 4.8).

Briegel (1990) has established in *Anopheles* spp. that up to 30% of the blood-meal protein (on a caloric basis) is utilised for the synthesis of yolk protein and lipid, and the number of eggs matured is closely correlated with the amount of blood consumed. Human blood generally has a higher protein content than rodent blood, but because of its low isoleucine content, only a limited amount can be channelled into oogenesis, thereby providing more protein for deposition in extra-ovarian tissues. This may be the reason why mosquitoes that are malaria vectors are attracted to human hosts. The metabolic requirement for isoleucine may, however, also explain why females in the field are forced to seek multiple blood meals during a single oogenic cycle. This, in turn, has most important implications in the spread of disease, particularly under crowded conditions when many poor people are obliged to sleep together in a single hut.

The sexual cycle in the female mosquito again affords us with a good example of the interplay between environmental stimuli, such as temperature, mating, nutrition, the recognition of favourable oviposition sites, and the endocrine system of this very successful group. In a similar fashion to vertebrate animals the central nervous system plays an essential integrating role.

4.1.3 Echinodermata

Echinoderms are exclusively marine invertebrates and have been classified into five major classes. The Asteroidea, or starfishes, are particularly suitable for laboratory studies and oocyte maturation, spawning and fertilisation have been very well studied in this class. The field has been extensively reviewed by Kanatani and Nagahama (1983).

Echinoderms are mostly bisexual but difficult to sex on the basis of their external appearance. In the starfishes each gonad consists of a group of elongated, tubular lobes and there are two gonads present in each arm giving a total of 10 gonads. When starfish reach maturity the ovaries usually have a bright orange colour because of the presence of yolk and the testes appear white to pale yellow due to very many sperm contained within them.

Most echinoderm eggs are fairly large, within the range 70–200 μm, and they are often enclosed in a gelatinous coating and a vitelline membrane. Yolking of the eggs can be accomplished by the oocyte itself, known as autosynthesis, and/or by heterosynthesis when a tissue other than the oocyte becomes involved in this process. In this regard it is most interesting to note that the normal female sex hormone oestradiol-17β, found in many vertebrates including the human female, causes rapid growth of the ovaries and a concurrent decrease in the size of the pyloric caecae in the starfish *Asterias rubens* (Schoenmakers *et al.*, 1978). It has also been confirmed that echinoderm ovaries contain various steroid hormones and the enzymes required for their synthesis. It is therefore not surprising that Schoenmakers *et al.* (1978) have also shown that the peak level of oestrone, close relative of oestradiol-17β, in the ovaries of *A. rubens* occurs at the same time as the pyloric caecae exhibit their highest protein content and just prior to the ovaries attaining their maximum protein content.

In most echinoderms spawning consists of releasing large numbers of gametes into the surrounding sea water. The number of eggs released can be incredibly large in some species (e.g. 100 million in *Asterias amurensis*), whereas brooding starfish will only shed 100–300 eggs. Induction of spawning is controlled first by a neurosecretory substance known as gonad stimulating substance (GSS), which has been extracted and purified from the radial nerves of the starfish *A. amurensis*, and finally by an ovarian hormone, 1-methyladenine (Kanatani, 1969).

Environmental control of gonad maturation and spawning appears to vary a great deal among the various classes of echinoderms. For example, Kanatani and Nagahama (1983) report that the Californian coast starfish, *Patiria miniata*, has an indefinite spawning time whereas the Japanese crinoid, *Comanthus japonica*, spawns only once in October between the first and last quarters of the moon. Because the mature gonads of sea-urchins are prized as a gourmet item in many Mediterranean countries, their appearance on local fish markets was seen to be correlated with the phases of the moon by classical Greek and Roman writers, including Aristotle. Professor Munro Fox provides a graphic description of how the lunar cycles from spring to September are closely correlated with the spawning cycle of sea-urchins in the Suez area (see Marshall, 1965). It has also been established that a rise in sea water temperature is important in inducing spawning and that cold summers postpone the breeding seasons of both sea-urchins and starfishes that normally breed in early autumn (Kanatani and Nagahama, 1983).

4.2 Patterns of reproduction

4.2.1 Asexual versus sexual reproduction

Sexual reproduction of living organisms is so commonplace that its significance seldom excites the interest of biologists beyond the physiological, cellular and behavioural mechanisms involved. Because of its universality, we assume that it has strong adaptive advantages and leave it at that. Yet the evolution of sexual reproduction still baffles the minds of many eminent evolutionists and remains a profound problem in that field. Space does not allow us to do justice to this field. Interested readers are referred to Stearns' (1987) review, *The Evolution of Sex and its Consequences*, for a thorough treatment of the subject.

4.2.2 Oviparity and viviparity

Animals that lay their eggs externally are described as oviparous. Those that retain their eggs until 'hatching' within the oviduct but provide them with sufficient nutrition in the form of yolk are usually termed ovo-viviparous, as they also give birth to live young. When the embryo becomes attached to the female reproductive tract and receives nutrition from the maternal circulation, the condition is known as viviparity. In some instances it is not very easy to make sharp distinctions between these reproductive patterns because, even among ovo-viviparous animals, an exchange of gases (O_2 and CO_2) must occur between the embryonic and maternal circulation. Also in some animals, particularly teleost fish, they may appear superficially to be ovo-viviparous whereas careful examination frequently reveals that they are viviparous. For example, Veith (1980) injected labelled amino acids into the maternal circulation of the inter-tidal fish, *Clinus superciliosus*, and using autoradiography traced the passage of the labelled amino acids through the ovaries and into the embryonic tissue. The embryos of this species are furnished with a minimum of yolk and their development is almost entirely dependent on the transport of nutrients from the maternal circulation to the embryo. These small fish inhabit the turbulent waters of inter-tidal pools exposed to strong surf action and it is not surprising that they have evolved away from the more primitive pattern of oviparity in these environmental circumstances.

Internal gestation is not confined to the vertebrates, and even some insects exhibit primitive forms of placentation, as does the curious arthropod *Peripatus edwardsii*. Certain species of shark develop a yolk-sac placenta and are truly viviparous. Some lizards and certain snakes are also known to produce live young from both yolk-sac and allantoic types of placenta. The only vertebrate class in which both ovo-viviparity and viviparity is entirely absent is the birds (Amoroso, 1964).

The obvious disadvantage of any form of viviparity is the limitation it places on the number of offspring that can be produced. We have already seen that certain echinoderms can produce millions of gametes, clearly an impossible feat if they were viviparous. On the other hand, the obvious advantage that viviparity confers is a much higher degree of protection of the offspring during early development and growth. It follows then that environmental influences must have had a very strong influence on the reproductive pattern adopted by particular organisms. Nevertheless, it is not a simple matter to pin-point a single environmental factor which has shaped the adoption of either viviparity or oviparity. For example, one could argue that an aquatic environment of great constancy would favour oviparity and the release of a large number of gametes by both sexes, as well as external fertilisation and embryonic development. Conversely, an unpredictable terrestrial environment would favour viviparity and a greater degree of parental care. These are reasonably safe generalisations to make but caution should be taken in identifying a single selection pressure or explanation for the development of either reproductive pattern. In the real world a complex of selection pressures is probably responsible and involves many factors concerned with the entire development, nutrition, growth, dispersal and predator evasion of the offspring. A simple example in one genus of lizard, namely *Chamaeleo*, should suffice to explain this point. *Chamaeleo namaquensis* is a ground-dwelling, desert lizard which inhabits extremely arid desert plains and dunes. During a single day it can exhibit a range in body temperatures from 2 °C in the early morning to 42 °C at midday. In view of the unpredictable and harsh environment which both the parents and offspring are exposed to, one would predict that this species would be viviparous or at least ovo-viviparous. In fact it is oviparous and lays its large eggs in hot sandy areas some 10 cm beneath the surface. In contrast, *Chamaeleo pumilus* lives in a temperate region amongst lush vegetation, cool ambient temperatures and high rainfall. Its body temperatures show a far smaller fluctuation than those of *C. namaquensis*. *Chamaeleo pumilus*, contrary to our generalisation, is ovo-viviparous (Burrage, 1973). There are several possible explanations for this apparent contradiction. The widely fluctuating body temperatures of the desert species may be unfavourable for normal embryonic development within the oviduct. Alternatively, the exclusively arboreal existence of the temperate species, *C. pumilus*, may preclude it from nest making and egg laying. Both explanations remain highly speculative, however, and illustrate how difficult it is to explain the evolution of these reproductive patterns. Nevertheless, their analysis is of crucial importance to the physiological ecologist and falls very much within our field of interest and responsibility.

4.2.3 Seasonal breeding

It is well known that almost all animals, except man, have a stated season for the propagation of their species. Thus the female cat receives the male in September, January and May. The she-wolf and fox in January; the doe in September and October. The spring and summer are the seasons appointed for the amours of birds, and many species of fishes. The immense tribe of insects have likewise a determinate time for perpetuating their kind; this is the fine part of the year, and particularly in autumn and spring. The last-mentioned class of beings is subject to a variation that is not observed in the others. Unusual warmth or cold does not retard or forward the conjunction of birds or quadrupeds; but a late spring delays the amours of insects, and an early one forwards them. Thus it is observed that, in the same country, the insects on the mountains are later than in the plains.

The above extract from Spallanzani's *Dissertations* (1784), quoted by Marshall (1965), although not entirely correct, shows that scientific interest i this important phenomenon is not new. In fact Aristotle discusses season: reproduction at some length and there can be no doubt that our hunte gatherer ancestors were well informed on the subject as they must have bec expert field biologists.

The evolution of seasonal breeding is obviously an adaptation to maximise the survival of offspring and for this reason it is synchronised in such a way that the offspring are produced at the most favourable time of the year. In the case of mammals, this means that sexual activity and mating must precede the most favourable time of the year by a period equivalent to the gestation period of a particular species. In most cases spring is the most favourable season and for this reason sheep, with a gestation length of 140 days, become sexually active in the autumn whereas horses, with a gestation length of 335 days, are most active in the spring. Although survival of offspring has probably been the most important selection pressure in the evolution of seasonal breeding, one should caution against over-emphasis of this factor because ultimately it is the survival of the species as a whole and not only the offspring which is controlled by natural selection. For example, in certain situations the survival of the offspring may have a low probability and survival of the parents at a particular point in time may be of far greater importance for the survival of the species. Moreover, each of the various phases of mammalian reproduction such as courtship, mating, gestation, parturition and lactation could hypothetically be as vulnerable to adverse environmental influences as the early life of the offspring. If a particular phase in the reproductive cycle is very vulnerable, then natural selection may be sufficiently strong to ensure that it occurs at the most favourable time for its particular requirements. Also, seasonal breeding may be entirely inappropriate in certain species of mammal with strongly organised social structures (Millar, 1972). In these species the sudden arrival of a large number of offspring, requiring parental care and imposing the heavy energetic demands of lactation simultaneously upon all the females, could disrupt the social structure very severely. An excellent comparative example in this regard is provided by the zebras and wildebeest that graze in very large numbers on the open savannahs of Africa. They are more or less the same size and, although the wildebeest are ruminants and the zebras monogastric, their nutrition is similar, consisting almost entirely of savannah grasses. They also live in exactly the same habitat but that is where the similarities end. The zebras are non-seasonal breeders, whereas the wildebeest show remarkable synchrony of breeding with nearly all the females producing calves within a matter of a few weeks. The mass production of young by the wildebeest saturates the environment with prey for its traditional predators such as lions, hyenas and wild dogs, and consequently the latter are unable to capture all the young calves and many survive to adulthood. In contrast to the huge wildebeest herds with their fairly weak social interactions, the plains zebras are organised into small family groups with a tight social structure. Within this structure the females undertake important sentinel duties to warn the group against imminent predation. If all the females were to foal simultaneously this social arrangement would probably break down and place the species at a much higher risk. The same arguments would also apply in the case of African wild dogs, baboon troops and perhaps humans as well. The ultimate example is that of eusocial animals, among which only one individual is permitted the luxury of reproduction at one time.

Skinner and Van Jaarsveld (1987) have examined the adaptive significance of restricted breeding in large southern African ruminants. They point out that the proximate cues controlling reproduction in these animals have only been established for very few species. However, their analysis does suggest that those species that are regular seasonal breeders are mostly found in the mesic, high-rainfall areas of the eastern part of the southern continent, whereas the opportunistic breeders occupy the western arid part where rainfall, via nutrition, appears to control reproduction. These authors also emphasise that the synchronisation of mating frequently depends on several subtle environmental cues and not just a single major influence. For example, in impala antelope the first matings occur within 6 days of the full moon in May, which coincides with a high incidence of male rutting calls during the full moon. The lunar cycle, in addition to the photoperiod, appears therefore to influence sexual activity in female impala, even if only secondarily through the behaviour of the males.

Similarly to the antelope of Africa, certain kangaroo species, such as the tammar wallaby, are strict seasonal breeders, whereas others, such as the red kangaroo, are opportunistic breeders, responding to sporadic rain in the arid regions of Australia.

The reproductive cycles of kangaroos are complex and have been extensively studied by Tyndale-Biscoe (1984) at the CSIRO Division of Wildlife Research near Canberra. His studies on female tammars have shown that they give birth to their undeveloped young during the hottest and driest time of the year. However, at this stage the immature joey, although attached to the teat within the pouch, makes minimal demands on the mother's nutrient reserves. When the joey begins to make heavy demands on the mother, the winter rains arrive and food is usually abundant. Figure 4.9 shows that within one day after giving birth to the immature joey, the mother exhibits behavioural oestrus. Should she mate successfully, the fertilised egg will only develop to the 80-cell blastocyst stage, and then remain in a dormant diapause stage for the next 11 months. During this period the suckling stimulus on the teat within the pouch stimulates the pituitary gland via neural pathways to secrete prolactin which, in turn, inhibits the growth of the corpus luteum, thereby preventing development of the blastocyst. Once the joey is weaned and leaves the pouch, this neural stimulus is ablated and the blastocyst is reactivated. Moreover, should the joey be lost or experimentally removed during the first half of the year, the corpus luteum will develop and the resulting progesterone will cause the blastocyst to develop, with the birth of an altricial joey occurring 27.5 days later. If the young joey is lost after the winter solstice (June in the southern hemisphere), reactivation of the blastocyst does not occur. Figure 4.10 illustrates the major hormonal influences

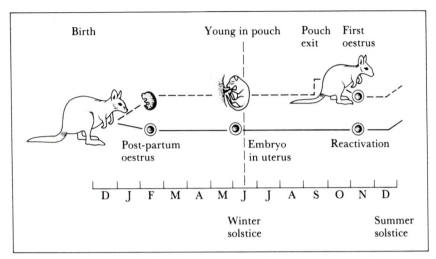

Figure 4.9   Seasonal breeding in the tammar wallaby begins with a post-partum oestrus in December when the young joey becomes attached to the teat within the pouch. At the same time the fertilised ovum remains in diapause due to the lack of progesterone for the following 11 months. Shortly after the mature joey leaves the pouch the quiescent blastocyst is reactivated and begins development. Maximum lactational demands of the developing joey are synchronised with winter rains and concurrent superior nutrition. After Tyndale-Biscoe (1982).

discussed above. Note that sexual activity in the male is not the result of direct photoperiodic stimulus upon an endogenous rhythm, but is rather induced by the pheromones and behaviour of the females. The complex cycle of the females has obvious adaptive value by ensuring that the maximum nutrient demands of the mother and the joey are synchronised with the most favourable season and, in the event of loss of the young, they can be immediately replaced by a new embryo kept in a state of diapause. The need for searching for a mate with all the attendant courtship rituals is also not required.

Delayed implantation of the blastocyst is also an important physiological adaptation in some eutherian mammals. It still allows birth of the young to take place at a favourable time, but mating need not precede parturition by a period of time equal to the gestation period. For example, a species of mammal with a short gestation period of say 8 weeks could mate under fairly favourable climatic conditions in the autumn when it is easier to seek out mates. Fertilisation takes place immediately but the blastocyst is not implanted and remains in diapause until 8 weeks before spring, whereupon implantation and development immediately begin. The best examples of this phenomenon are found among the Mustelidae (e.g. North American pine

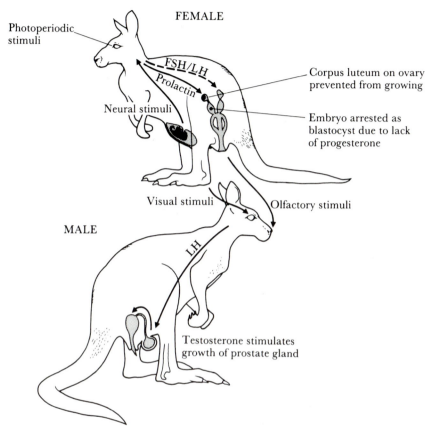

Photoperiodic stimuli

FEMALE

FSH/LH

Prolactin

Corpus luteum on ovary prevented from growing

Neural stimuli

Embryo arrested as blastocyst due to lack of progesterone

Visual stimuli

Olfactory stimuli

MALE

LH

Testosterone stimulates growth of prostate gland

Figure 4.10 Interplay of hormones in the tammar wallaby which suppresses development of the embryo, while stimulating milk secretion for the developing joey in the pouch. Note that the male wallaby is not stimulated directly by the photoperiod but rather by pheromones and optical stimuli emanating from the female. After Tyndale-Biscoe (1982).

marten). Certain seal species also show delayed implantation which allows them to combine parturition with the mating season during a single period spent on land, known as the annual 'haul out'. Most invertebrates, the lower vertebrates and small mammals can proceed from the initiation of sexual activity to the production of young so rapidly that the above adaptations are unnecessary and they usually become sexually active immediately upon the advent of the most favourable time of the year.

Possible delays in finding suitable mates may occur, however, and to provide for this, many female insects and reptiles have evolved physiological mechanisms for storing sperm in a dormant condition. When favourable climatic conditions prevail, the sperm is activated and fertilisation and

Plate 4.1   A Ross seal photographed on an Antarctic ice floe. It is protected from
the cold by a thick specialised pelage and a layer of blubber, which forms
a cold shell around the animal. Seals also exhibit delayed implantation of
the fertilised ova. This allows them to give birth to their pups at the most
favourable time of the year, without having to mate on a date which
precedes parturition by the true gestation period. Many seal species
mate shortly after parturition during the annual 'haul out'. (Photo:
courtesy J. D. Skinner.)

embryonic development ensue immediately. Chameleons are a good example
of female vertebrates that store sperm for several years, producing several
litters from a single mating. Certain bird species are able to store sperm but
for much shorter periods.

The initiation of sexual activity at the appropriate season depends on
suitable environmental cues impinging on the organism at the correct time.
These cues are often referred to as proximate environmental factors and
several may be involved for the completion of a full reproductive cycle. For
example, small desert mammals may respond initially to an increasing
photoperiod in the spring, if that is the season with the highest probability for
rainfall to occur. They would, however, not become fully active sexually until
a certain threshold amount of rain had fallen to produce the growth of
protein-rich seedlings. By remaining at sub-threshold activity until the rain
falls they are conserving their resources for optimal breeding conditions.

The sexual response of amphibians to either rising temperatures in spring
in mesic areas or rain in arid environments is well known; the males begin to
call vociferously, amplexus occurs in the water and the larval tadpoles

usually develop in water while feeding on microscopic food particles. There are, however, interesting exceptions to this standard pattern. The arid-adapted African frog *Chiromantis* lays its eggs in a foam nest on a branch suspended over a temporary water pool. When the eggs hatch the larvae fall into the pond and begin feeding. Even more unusual is the Australian gastric breeding frog *Rheobatrachus silus*. It is able to convert its stomach into a brood pouch for 6–7 weeks. During this period gastric secretion of acid is inhibited by a prostaglandin and eventually the young (as many as 26) are born orally as perfectly formed miniature frogs (Tyler, 1985).

The photoperiod, or changing daylight length, is one of the most widely adopted environmental cues for initiating sexual activity in animals and many plants. The reason for this is that it is entirely dependable and never changes from year to year. Temperature, nutrition, optical and auditory stimuli are also important proximate factors which can act either as the primary stimulus to entrain the reproductive cycle or as secondary or tertiary stimuli at different phases of the cycle. Lunar cycles are important in certain species, as we have seen in the echinoderms and even the sight of rainfall alone can act as the final stimulus for nest building and egg laying in desert birds.

Seasonal breeding is not always accompanied by synchronous maximum development of the gonads, as illustrated in the following example. The red-sided garter snake (*Thamnophis sirtalis parietalis*) successfully inhabits and reproduces under the extremely cold conditions of western Canada, where winter temperatures frequently fall below $-40\,°C$ and the snow cover is often permanent throughout winter. Its distribution is the farthest north of any reptile in the western hemisphere. How does it succeed under these extreme conditions? Throughout the winter they shelter in large communal sinkholes, far beneath the surface. The dens or caverns within these sinkholes can house as many as 10000 snakes. In spring the males emerge first and almost simultaneously. The resulting, writhing mass of snakes is so dramatic that the phenomenon has become an important tourist attraction. The females then emerge from the den singly or in small groups. They are immediately set up by the males and the great preponderance of males means that as many as 100 males at a time are attempting to mate with a single female. This writhing mass is known as a mating ball (Fig. 4.11) and usually females have been successfully mated within 30 min of leaving the sinkhole den.

Once the females have mated they disperse from the sinkhole to begin feeding in the surrounding prairie vegetation. The mating period lasts between 3 days and 3 weeks, whereafter the males also disperse to the feeding areas. The young are born alive and as many as 30 can be produced by a single female. The adults return to the dens to hibernate during early autumn and the entire cycle has to be completed within only three summer months. However, unlike many other seasonal breeders, in which the reproductive organs begin to develop in late winter so as to be at peak activity in

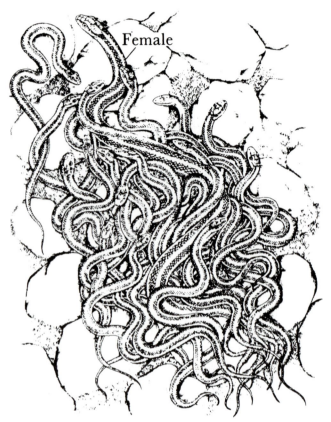

Figure 4.11   When the red-sided garter snakes of western Canada emerge from their underground dens in spring, the males far outnumber the females and their intensive pursuit of the few females, during this period, results in the formation of a writhing mass of snakes known as a 'mating ball'. Redrawn from Crews and Garstaka (1982).

spring, the garter snakes emerge in spring when their gonads are at their smallest and the levels of circulating sex hormones are at their lowest. David Crews and William Garstaka (1982) have studied the garter snake extensively and describe this phenomenon as a mismatch between the behaviour and the physiological status of the snakes. In spite of extensive experimentation with various gonadal and pituitary hormones, these authors are still not able to explain which physiological trigger is responsible for the intense mating behaviour of the males in spring when, theoretically, the low blood levels of androgens should result in a state of very low sexual activity.

Figure 4.12 illustrates the reproductive cycle of both male and female garter snakes. From January until mid-May the snakes are hibernating in the sinkholes. In both females and males blood concentrations of the sex hor-

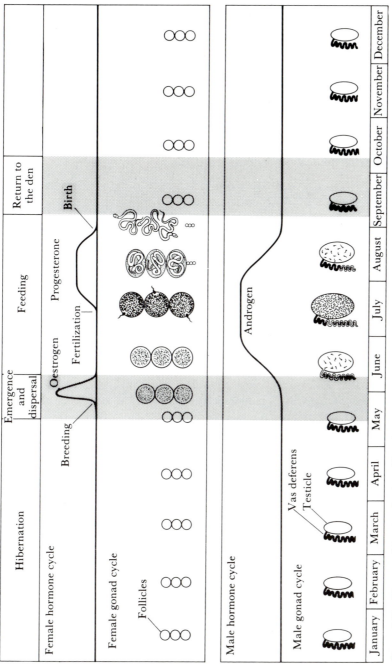

Figure 4.12 The gonadal cycle of male and female red-sided garter snakes. Note the mismatch between hormonal levels and sexual behaviour. See text for further details. Redrawn from Crews and Garstaka (1982).

mones are low and the gonads are undeveloped but the male's vas deferens is filled with stored sperm. Mating occurs in late May and very soon thereafter oestrogen levels peak rapidly in the female and androgens begin to rise after the mating period. The increase in oestrogen levels in the female stimulate the ova to develop and become yolked. Stored sperm is then activated towards the middle of July and the ova are fertilised. Progesterone blood levels begin to rise until parturition occurs in August or early September. In the male the rising levels of androgens in summer are accompanied by growth of the testes and sperm production. The sperm are stored in the epididymis for use during the next spring. Towards the end of September both male and female adults will have returned to their dens. Little is known about the overwintering of the juveniles born in August.

4.2.4 Precocial and altricial breeding

The production of semi-helpless (altricial) young at birth or hatching, as opposed to fully developed precocious young, has attracted considerable interest from evolutionary biologists who have attempted to explain the adaptive value of these modes of reproduction.

Ricklefs (1974) believes that in birds natural selection should favour precocity, because independent feeding by the chicks should theoretically increase the number of chicks that can be raised, if all other environmental factors are equal. Ar *et al.* (1987) have examined the energy in precocial and altricial avian eggs and hatchlings and conclude that in spite of differences in total energy content, the energy densities $(kJ \ g^{-1})$ of dry matter in the eggs and the efficiencies with which energy is transformed from eggs to hatchlings are independent of maturity at hatching. They did find, however, that the main differences among maturity types were different water concentrations of eggs and hatchlings, differing amounts of energy transported in an untransformed state from the egg to spare yolk and in the concentration of energy in the dry matter of hatchlings. Vleck and Vleck (1987) conclude from their studies that altricial species have lower costs of development, smaller hatchlings and shorter incubation periods than precocial birds when using egg mass as a scaling variable. Nevertheless, they arrive at the same conclusion as Ar *et al.* (1987) in stating that altricial and precocial species do not apparently differ in the efficiency with which energy is transformed into yolk-free dry matter contained in the hatchling $(15.4 \ kJ \ g^{-1})$. Vleck and Vleck (1987) suggest that during the evolution of altricial birds from their precocial ancestors the first step was the reduction of the incubation period. This, in turn, reduced the amount of energy required before hatching and, as a result, the female could reduce the amount of energy invested per egg by decreasing its size and by increasing its water content. It would seem then that the energetic cost of development in altricial and precocial birds does not differ significantly. However, precocial birds produce eggs with more total energy and less water than altricial species which, on the other hand, are obliged to invest more energy in supplying food to their altricial young during the post-hatching period.

The ecological and evolutionary implications of these modes of reproduction are complex and simple explanations are seldom appropriate. Nevertheless, the primary importance of the food resources available to the chicks must be emphasised. Precocial chicks require food that can be easily obtained such as seeds, young vegetation or sluggish insect larvae. Prey items that require strength, experience and skill to capture, such as flying insects, require adult predators which then feed their altricial young in protected nests. For an extensive treatment of the subject see Ar and Yom-Tov (1978) and O'Connor (1984).

In the case of mammals there is also not a great difference between the energetic cost of raising precocial or altricial young, although there may be a slightly higher energetic cost when providing nutrition to the young by lactation, as opposed to placental nutrition *in utero*, because the former process requires additional synthesis of nutrients. The difference is not critical and other factors such as the availability of protected nesting sites in the natural habitat, body size and the nature and distance of food resources are far more important in determining the mode of reproduction. Altricial breeding in mammals does, however, have one obvious advantage over precocial breeding in that the uterus can be vacated fairly soon after conception to allow immediate replacement with a second litter, while the first litter is still being suckled. This naturally places a great energetic demand on the mother but speeds up the reproductive rate, which would be a great advantage to, say, a small desert rodent that has to contend with a very brief season of favourable nutrition after a rainfall event. Large mammals, on the other hand, such as antelope cannot retire to a safe den and, consequently, produce remarkably precocial young which are able to sprint away from predators within a matter of minutes after birth. The felids, canids and primates, including humans, can retire to safe protected dens and resting areas and produce fairly altricial young. This may enhance the parent–offspring bond, thereby facilitating the learning processes which these animals have to undergo.

4.2.5 Reproductive success

Repeated reproductive success of a species depends not only on how efficiently an organism reproduces itself. It also depends on the survival of the progeny to the reproductive stage and on the survival of the parents in some cases. For this reason the theoretical argument that semelparous animals, i.e. organisms that reproduce once and then die, should reproduce far more efficiently than iteroparous species, those that reproduce repeatedly, is not valid under natural conditions. As Callow (1978) has pointed out, iteroparity provides an insurance against the mass loss of progeny at any one time and it is surprising that it has not evolved more frequently. He concludes that the most successful theoretical reproductive pattern should be a long lifespan, high reproductive output and repeated breeding, but because energy and nutrients must be used for a host of activities, other than reproduction, this pattern is rare, e.g. endoparasites. In fact, an inverse relationship between reproductive output and repeated breeding is frequently found within taxa (Callow, 1978).

These considerations of the advantages of semelparity over iteroparity lead us to an examination of some of the quantitative aspects of reproduction. Why, for example, do certain species produce hundreds of thousands of offspring with no parental care, whereas others produce a single offspring and invest heavily in parental care? Are animals reproducing as swiftly as possible within the constraints imposed upon them by their biology and their environment, as suggested by Lack (1948) or are they breeding as slowly as they can as Baker (1938) proposed? These questions led to the concept of r- and k-selected species proposed by MacArthur and Wilson (1967). Pianka (1970) has listed several important correlates of r- and k-selection. His summary shows that r-selected species produce a large number of offspring and that this pattern is usually associated with an unstable environment, unpredictable resources, catastrophic mortality, variable population size, rapid development, a short lifespan and small body size, whereas k-selection is usually correlated with the opposite conditions and characteristics. This concept provides a useful framework for evaluating reproductive patterns but because few species fit perfectly into the two extreme categories, Pianka (1970) has suggested that we should rather visualise an r–k continuum and evaluate each case separately.

For a recent and thorough review on reproductive success, the interested reader is referred to Clutton-Brock (1988). This review includes many examples from a wide variety of taxa. In his summary of this volume, Clutton-Brock examines the question of whether variation in breeding success occurs by chance. This is best answered by investigating if variations in phenotype are related to differences in breeding success. If they are, then the commonly observed variation in mating success cannot be due to random variation in access to mates. Based on this criterion, Clutton-Brock (1988) concludes that chance seldom accounts for the measured variation in breeding success and that the most important phenotypic characteristics related to breeding success in both males and females are age, dominance rank, body size, mate choice and early development. In addition, several environmental factors have a significant effect on breeding success, e.g. climatic variation, local differences in habitat quality, changes in population density and, in the case of social species, group size and changes in group membership (Clutton-Brock, 1988).

4.3 Evolutionary enigma of the scrotum and sperm competition

Although there are many exceptions, in most mammalian families the male gonads or testes are carried outside the abdominal cavity in the scrotum. In ancient times, members of certain human tribes would lay their hands on the scrota of other members as an act of faith or to seal an agreement. Hence the origin of present-day terms such as 'testify' and 'testament'. It is also possible that the ancient scholars pondered on the enigma of the evolution of the scrotum. Why, for example, should the most important organs involved in reproducing the species be carried in such a vulnerable position? Several hypotheses have been advanced to explain this phenomenon and the first to be debated was based on the findings of Crew (1922). Crew was wounded in the First World War and as a result retained a scrotal fistula, which allowed him to measure the temperature of his testis and scrotum under various conditions. On the basis of his findings it was established that the testes are held at a temperature some 4–5 °C lower than normal body temperature and that under hot conditions the cremaster muscle relaxes and the testes are moved further away from the abdomen. Under cool conditions the converse is true. In addition to the contraction and relaxation of the cremaster muscle, cooling of the testes is further facilitated by the presence of the pampiniform plexus (Fig. 4.13) which, similar to the carotid rete discussed previously, acts as a countercurrent heat exchanger. The high density of sweat glands on the surface of the scrotum ensures that the venous blood is cooled efficiently before returning to the coiled arteries in the pampiniform plexus where the heat is exchanged. It seems therefore that many mammals go to rather extraordinary lengths to cool their brains and testes. Later it was shown that if the testes of scrotal mammals are transplanted into the abdominal cavity, spermatogenesis soon ceases but testosterone secretion remains almost normal. Similarly, when temperate breeds of cattle are exported to the tropics, the high ambient temperatures cause a sharp reduction in semen quality with many abnormal sperm, whereas the indigenous cattle breeds of the *Bos indicus* type are far less affected by high temperatures (Skinner and Louw, 1966). These findings led to the belief that scrotal testes evolved because spermatogenesis required a temperature lower than body temperature to proceed optimally. Let us examine this and some other hypotheses more critically.

First, we can dispense with the temperature/spermatogenesis hypothesis very easily by pointing out that spermatogenesis occurs quite normally in so-called testicond mammals, in which the testes are carried within the abdominal cavity. Moreover, spermatogenesis in birds, with body temperatures as high as 42 °C, also proceeds normally. A second hypothesis holds that the pampiniform plexus in scrotal mammals performs a so-called haemodynamic function by dampening the arterial pulse, thereby protecting the sensitive germinal epithelium and spermatogenic process from wide fluctuations in

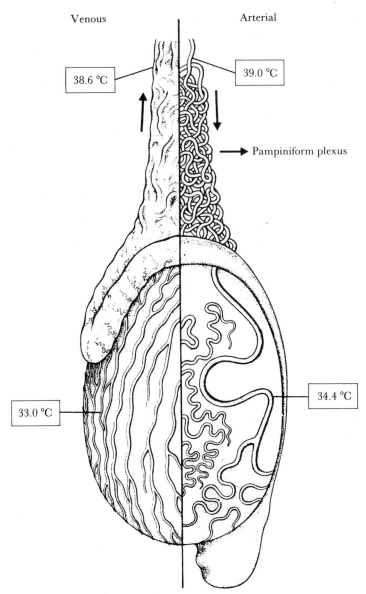

Venous   Arterial

38.6 °C   39.0 °C

→ Pampiniform plexus

34.4 °C

33.0 °C

Figure 4.13 The position of the pampiniform plexus in the mammalian testis, which acts as a countercurrent heat exchanger to keep the temperature of the testis and epididymis over 4 °C lower than core temperature. After Setchell (1982).

blood pressure. This argument can again be refuted by the fact that spermatogenesis in testicond mammals and birds proceeds normally without the intervention of the pampiniform plexus. It has also been argued that the scrotum has developed as a sexual signal to females and proponents of this hypothesis point to the highly coloured scrota of some primates as the best example of this phenomenon. Although it is quite probable that the scrotum could function as an optical releasing or trigger mechanism in certain primate species, this hypothesis cannot be accepted as a broad generalisation because the scrota of most mammalian species are usually rather cryptic. Finally, it has been suggested that the scrotum has not evolved to allow spermatogenesis to proceed at a lower temperature but rather to store sperm in the epididymides at a lower temperature, thereby reducing the metabolic rate of the sperm, possibly reducing their mutation rate as well, and prolonging their lifespan. It follows then that this advantage would be particularly important to promiscuous males that copulate frequently and that they would exhibit the best-developed scrota and testes. This seems to be the case when we compare the testicond mammals, Proboscidae (elephant), Hyracoidea (hyrax), Sirenia (dugong), Macroscelididae (elephant shrews), Chrysochloridae (golden moles), Tenrecinae (tenrecs), Monotremata (duck-billed platypus), Edentata (sloths), Cetacea (whales), and members of the Oryzorictinae and Solenodontidae, with all the remaining mammalian families that possess scrota. In this regard it is particularly significant that the ruminants and primates have very well-developed scrota and are generally the most promiscuous. Although the evolution of the scrotum remains an enigma, this final suggestion is the most convincing.

4.4 Lactation

Lactation is defined as the secretion of nutrients by mammary glands and is therefore restricted to the vertebrate class Mammalia. Nevertheless, a very similar process occurs when crop 'milk' is secreted by certain bird species.

Mammary glands have evolved from pseudiferous (sweat) glands and during the earliest phase of their evolution they were probably employed to moisten the eggs of egg-laying mammals. Mammary glands undergo rapid development towards the end of gestation and in most species regress after lactation ceases. In humans, however, the mammaries remain prominent throughout adult life. This has led to considerable speculation on their alleged role as intensifiers of the pair bond in humans, the latter being of critical ecological importance to hunters and gatherers.

The hormonal control of mammary gland development and lactogenesis has been discussed previously (see sexual cycle of the ewe). Figure 4.14 provides more detail in this regard and shows that, in addition to oestrogens, progesterone and prolactin, both growth hormone and adrenal steroids are required for full development of a functional mammary gland. The major hormone involved in the actual initiation of milk secretion is prolactin. It is a complex protein secreted by the anterior pituitary gland and has an interesting evolutionary history. It is present in all vertebrates except cyclostomes

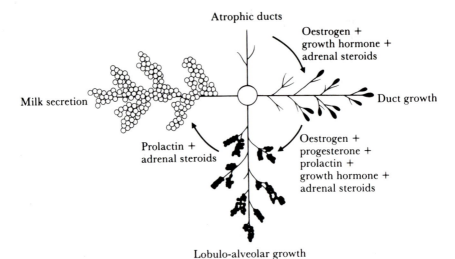

Atrophic ducts

Oestrogen +
growth hormone +
adrenal steroids

Milk secretion

Duct growth

Prolactin +
adrenal steroids

Oestrogen +
progesterone +
prolactin +
growth hormone +
adrenal steroids

Lobulo-alveolar growth

Figure 4.14   The major hormones involved in the growth and development of the mammary gland of the laboratory rat after removal of the pituitary and adrenal glands.

and fulfils a variety of functions, such as the 'water drive' in breeding sala-manders, and affecting gill function and glomerular filtration rate in certain teleosts as well as crop milk production in pigeons. Of perhaps even greater interest is its ability to stimulate mucus secretion on the skin of certain fish. In one such fish species (*Symphysodon*) the newly hatched young have been observed to feed off the skin surface of the parental fish, suggesting another, more ancient lactational-like role for prolactin. In the latter respect it should be remembered that the mammary glands are modified skin glands.

The composition of milk varies greatly among species. It always contains water, lipids, carbohydrates, proteins (mostly casein) as well as minerals and vitamins. It is, however, frequently deficient in iron. The variation in milk composition of mammals is illustrated in Table 4.1, which shows the much higher percentage of fat and protein in the milk of marine mammals. This has been ascribed to the need for water conservation in the marine environ-ment, but a more convincing explanation is the high requirement of energy by the neonatal marine mammal. They must not only grow rapidly but also deposit large amounts of blubber for thermal insulation and buoyancy. Another interesting difference in milk composition, not included in Table 4.1, is the energy content of rhinoceros milk ($500\,\text{kcal}\,\text{kg}^{-1}$) compared with reindeer milk ($2773\,\text{kcal}\,\text{kg}^{-1}$). Rhinos live in the hot African savannah and the high water content of their milk probably compensates for high rates of evaporative water loss from the juvenile rhinos. In contrast, reindeer are exposed to subzero temperatures and the young must acquire a well-

Table 4.1. Variation in milk composition of selected mammals

Species	Fat (%)	Protein (%)
Rhinoceros	0.3	1.2
Human	3.8	1.2
Cow	3.7	3.4
Horse	1.9	2.5
Blue whale	42.3	10.9
Grey seal	53.2	11.2

developed layer of subcutaneous fat to survive in the sub-Arctic environment.

Before leaving the subject of milk composition, the importance of colostrum to neonatal mammals should be noted. Colostrum is the milk produced for a few days immediately after parturition. It is rich in globulins and confers passive immunity to certain diseases in the neonate. At a later stage the young animals must naturally develop their own active immunity to these diseases. Colostrum is particularly important in ruminant animals because the placenta in these animals (syndesmochorial) does not allow the free transport of maternal antibodies to the foetal circulation.

Finally, we must examine the adaptive or survival value of lactation in an ecological/evolutionary context. Many advantages could be cited and it is a rich field for speculation. Here are just a few possible advantages.

1. Lactation promotes the development of a strong bond between parent and offspring. This is particularly important for mammals that depend heavily on learned behaviour, e.g. primates, felids and canids.

2. In the case of ruminants and caecal digesters, very coarse fibrous plants with a low nutritional value can, with the aid of symbiotic organisms in the digestive tract and specialised biochemical mechanisms in the mammary gland, be transformed into a highly nutritious and digestible fluid, namely milk.

3. Lactation allows the delivery of a relatively large quantity of food to neonates within a short period when necessary. This allows absence from the nest or young for fairly long periods while foraging. For example, some marine mammals only suckle once per week.

4. Lactation allows the birth of a relatively large number of altricial young with underdeveloped jaws. These neonates can be immediately replaced with another full complement of foetuses in the uterus. Gestation therefore proceeds concurrently with lactation and the rate of reproduction during a short favourable season is maximised. This would, for example, be of great advantage to an *r*-selected species.

5. Long before lactation commences, a female mammal could store nutrients in her tissues during a time of plenty. These could be catabolised

(a)

(b)

Plate 4.2　The rich milk of the reindeer (a) has a calorific value of 2773 kcal kg^{-1}, whereas the milk of the rhinoceros (b) has an energy value of only 500 kcal kg^{-1}. The dilute milk of the rhino is probably an adaptation to the hot African savannah so that evaporative water loss from the baby rhino can be readily replaced. The opposite probably holds true for the reindeer in which rapid fat deposition is of greater importance. (Photo (a): courtesy N. J. C. Tyler.)

during lactation to feed her young, even if environmental conditions at that time were unfavourable. Birds do not enjoy the same advantage when feeding their altricial nestlings and a comparison of birds and mammals in this respect would be of considerable ecological interest.

4.5 Concluding remarks

As pointed out at the beginning of this chapter the ultimate 'fitness' of any species is measured by its reproductive success. The reproductive success, in turn, is influenced by the interaction of the species with countless environmental factors such as nutrition, temperature, water availability, parasites, predation and the availability of a suitable habitat, to name but a few. It is therefore not surprising that a myriad of reproductive patterns has evolved in the animal kingdom in response to these selection pressures. These include internal and external fertilisation, semelparity and iteroparity, altricial and precocial reproduction, oviparity and viviparity, sexual and asexual reproduction, seasonal breeding, sperm storage and lactation.

In this chapter we have examined these patterns in an attempt to relate them to their survival value for the species concerned. The overriding conclusion that can be drawn from this analysis is very similar to the conclusions drawn from the preceding chapters, namely that simplistic explanations for a particular species' reproductive pattern are seldom appropriate and that reproductive patterns are a reflection of a complex of selection pressures in that species' evolutionary history. For example, a particular desert rodent may live in a burrow to escape the heat and desiccating conditions on the surface of the desert. By living in a protected burrow, the animal can produce large numbers of altricial young, thereby exploiting a short favourable season of plenty after a rare rainfall event. Being a mammal, it can provide highly nutritious milk to the altricial young, by breaking down its own body reserves if necessary, while the next litter is developing *in utero*. Its small size means that it is not able to locomote efficiently and must by necessity survive on the food resources in its immediate vicinity. Also, the substrate must be suitable for easy burrowing. Other desert rodents living, for example, on rocky hillsides are unable to burrow and consequently produce fewer precocial young. In contrast, large desert antelope such as the oryx produce precocial young, independent of the arrival of rain. The calves can sprint away from predators within minutes of being born and both the young and the parents can move efficiently over long distances in search of food and water. In contrast, the minute tadpole shrimps (notostracans) survive in the form of desiccated eggs on the desert surface for years on end and in a state of suspended animation, and are able to survive temperatures as high as 98 °C. The arrival of a threshold amount of rain immediately revitalises these eggs and the adult develops and proceeds to produce several thousand eggs in her short lifetime. This is an excellent example of *r*-selection with a short lifespan

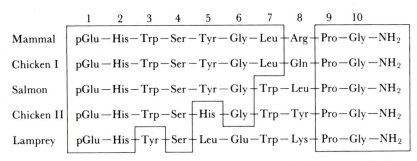

Figure 4.15 Primary structures of gonadotrophin-releasing hormones (GnRHs) isolated from vertebrate brain. Blocked areas show conserved regions. After Millar and King (1988).

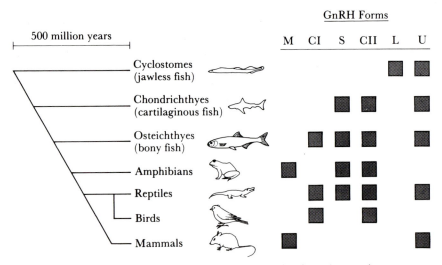

Figure 4.16 Phylogenetic distribution of gonadotrophin-releasing hormones (GnRHs) from over 30 species. GnRHs are designated M (mammalian), CI (chicken I), S (salmon), CII (chicken II), L (lamprey), and U (unidentified forms). From Millar and King (1988).

and high reproductive output in a very unstable environment. These minute crustaceans that inhabit ephemeral desert ponds usually reproduce by parthenogenesis as long as conditions are favourable but, as conditions begin to deteriorate, males are produced and the eggs produced from sexual reproduction have superior resistance to desiccation and are therefore suitable for the next long diapause stage.

These few examples, together with all those discussed previously, should be adequate to emphasise the great variety of reproductive patterns that exist among animals and how intimately they are interwoven with environmental influences. Students requiring a more detailed knowledge of reproductive physiology are referred to a very readable series edited by Austin and Short (1972). However, in spite of this great variety of reproductive patterns it is important to note that at the cellular level the process of fertilisation, the action of sex hormones and the interaction between hormones and the central nervous system are remarkably similar over a wide range of taxa. We have seen that the oestrogen-like hormones produced by starfish are almost identical to those employed by human females. Moreover, Millar and King (1988) have unravelled the amino acid sequence of the well-known decapeptide GnRH in the lamprey, chicken, salmon and mammal and their results show remarkable 'conservation' of the basic structure of the hormone during evolution of the vertebrates (Fig. 4.15). Their results also show an interesting distribution of the various types of GnRH among modern vertebrates (Fig. 4.16).

The wide divergence of reproductive patterns and the similarity of cellular processes in reproductive physiology is not surprising in view of the ancient and common cellular ancestry of all animals. There are many similar comparisons. As pointed out previously, the prehension of food by the mosquito and the elephant are very different, whereas digestion of their food has certain similarities but metabolism of nutrients within their cells is almost identical.

Bibliography

Adams, N. J., Brown, C. R. and Nagy, K. A. (1986) Expenditure of free-ranging wandering albatrosses *Diomedea exulans*. *Physiol. Zool.* **59**: 583–91.

Adams, N. J., Abrams, R. W., Siegfried, W. R., Nagy, K. A. and Kaplan, I. R. (1991) Energy expenditure and food consumption by breeding Cape gannets *Morus capensis*. *Mar. Ecol. Prog. Ser.* **70**: 1–9.

Ahearn, G. A. (1970) The control of water loss in desert tenebrionid beetles. *J. Exp. Biol.* **53**: 573–95.

Ahlquist, D. A., Nelson, R. A., Steiger, D. L., Jones, J. D. and Ellefson, R. D. (1984) Glycerol metabolism in the hibernating black bear. *J. Comp. Physiol. B* **155**: 75–9.

Akhmerov, R. N. (1986) Qualitative difference in mitochondria of endothermic and ectothermic animals. *FEBS Lett.* **198**: 251–5.

Alexander, R. M. (1982) *Locomotion of Animals.* Blackie, London.

Amanova, M. (1975) Adaptive features of the relief structure of the intestinal mucous membrane in desert birds. *Dokl. Akad. Nauk. SSSR* 225–510. Quoted by Phillips, J. G., Butler, P. J. and Sharp, P. J. (1985) *Physiological Strategies in Avian Biology.* Blackie, London.

Amoroso, E. C. (1964) Placentation. In *Marshall's Physiology of Reproduction*, Vol. II (ed. Parkes, A. S.). Longman, London.

Ar, A. and Yom-Tov, Y. (1978) The evolution of parental care in birds. *Evolution* **32**: 655–69.

Ar, A., Arieli, B., Belinsky, A. and Yom-Tov, Y. (1987) Energy in avian eggs and hatchlings: utilization and transfer. *J. exp. Zool.* (Suppl. 1): 151–64.

Arp, A. J. and Childress, J. J. (1983) Sulfide binding by the blood of the hydrothermal vent tube worm *Riftia pachyptila*. *Science* **219**: 295–7.

Austin, C. R. and Short, R. V. (1972) *Reproductive Patterns.* Cambridge University Press, Cambridge.

Baker, H. G. and Baker, I. (1982) Floral nectar sugar constituents in relation to pollinator type. In: *Handbook of Experimental Pollination Biology* (eds Jones, C. E. and Little, R. J.). Van Nostrand Reinhold, New York.

Baker, J. R. (1938) The evolution of breeding seasons. In: *Evolution, Essays on Aspects of Evolutionary Biology* (ed. Goodrich, E. S.). Clarendon, Oxford.

Baker, M. A. and Hayward, J. N. (1967) Carotid rete and brain temperature of the cat. *Nature* **216**: 139–41.

Bakken, G. S. (1976) An improved method for determining thermal conductance and equilibrium body temperature with cooling curve experiments. *J. Thermal Biol.* **1**: 169–75.

Bakken, G. S. (1980) The use of standard operative temperature in the study of the thermal energetics of birds. *Physiol. Zool.* **53**: 108–19.

Bakker, R. T. (1972) Anatomical and ecological evidence of endothermy in dinosaurs. *Nature* **238**: 81–5.

Barnett, S. A. and Little, M. J. (1965) Maternal performance in mice at −3 °C: food consumption and fertility. *Proc. Roy. Soc. B* **162**: 492–501.

Bartholomew, G. A. (1982) Body temperature and energy metabolism. In: *Animal Physiology, Principles and Adaptations*, 4th edn (ed. Gordon, M. S.). Macmillan, New York.

Bartholomew, G. A. and Lasiewski, R. C. (1965) Heating and cooling rates, heart rate and simulated diving in the Galápagos marine iguana. *Comp. Biochem. Physiol.* **16**: 573–82.

Bartholomew, G. A. and Lighton, J. R. B. (1986) Oxygen consumption during hover-feeding in free-ranging Anna hummingbirds. *J. exp. Biol.* **123**: 191–9.

Bartholomew, G. A., White, F. N. and Howell, T. R. (1976) The thermal significance of the nest of the sociable weaver *Philetairus socius*: summer observations. *Ibis* **118**: 402–10.

Bartholomew, G. A., Lighton, J. R. B. and Louw, G. N. (1985) Energetics of locomotion and patterns of respiration in tenebrionid beetles from the Namib Desert. *J. Comp. Physiol. B.* **155**: 155–62.

Bartholomew, G. A., Lighton, J. R. B. and Feener, D. H. Jr (1988) Energetics of trail running, load carriage, and emigration in the column-raiding army ant *Eciton hamatum*. *Physiol. Zool.* **61**: 57–68.

Bartlett, P. N. and Gates, D. M. (1967) The energy budget of a lizard on a tree trunk. *Ecology* **48**: 315–22.

Benedict, F. G. (1938) *Vital Energetics: A Study in Comparative Basal Metabolism*. Carnegie Institute of Washington, Washington DC.

Bennett, A. F. (1980) The thermal dependence of lizard behaviour. *Anim. Behav.* **28**: 752–62.

Bennett, A. F. (1987) Evolution of the control of body temperature: is warmer better? In: *Comparative Physiology: Life in Water and on Land* (eds Dejours, P., Bolis, L., Taylor, C. R. and Weibel, E. R.). IX-Liviana Press, Padova.

Bennett, A. F. and Ruben, J. A. (1979) Endothermy and activity in vertebrates. *Science* **206**: 649–54.

Bennett, A. F. and Ruben, J. A. (1986) The metabolic and thermoregulatory status of therapsids. In: *The Ecology and Biology of Mammal-like Reptiles* (eds Hotton, N. III, MacLean, P. D., Roth, J. J. and Roth, E. C.). Smithsonian Institution Press, Washington DC.

Bentley, P. J. (1960) Evaporative water loss and temperature regulation in the marsupial *Setonyx brachyurus*. *Aust. J. Exp. Biol. Med. Sci.* **38**: 301–6.

Bernays, E., Edgar, J. A. and Rothschild, M. (1977) Pyrrolizidine alkaloids sequestered and stored by the aposematic grasshopper, *Zonocerus variegatus*. *J. Zool. Lond.* **182**: 85–7.

Bertsch, A. (1984) Foraging in male bumblebees (*Bombus lucorum* L.): maximising energy or minimising water load? *Oecologia* **62**: 325–36.

Blaylock, L. A., Ruibal, R. and Platt-Aloia, K. (1976) Skin structure and wiping behavior of phyllomedusine frogs. *Copeia* 1976: 283–95.

Bligh, J. (1973) *Temperature Regulation in Mammals and Other Vertebrates*. North-Holland, Amsterdam.

Bligh, J., Louw, G. N. and Young, B. A. (1976) Effect of cerebro-ventricular administration of noradrenaline and carbachol on behavioural and autonomic thermoregulation in the monitor lizard *Varanus albigularis albigularis*. *J. Thermal Biol*. **1**: 241–3.

Blix, A. S. and Johnsen, H. K. (1983) Aspects of nasal heat exchange in resting reindeer. *J. Physiol*. **340**: 445–54.

Blix, A. S., Grav, H. J., Markussen, K. A. and White, R. G. (1984) Modes of thermal protection in newborn muskoxen (*Ovibos moschatus*). *Acta Physiol. Scand*. **122**: 443–58.

Block, B. A. (1987) Strategies for regulating brain and eye temperatures: a thermogenic tissue in fish. In: *Comparative Physiology: Life in Water and on Land* (eds Dejours, P., Bolis, L., Taylor, C. R. and Weibel, E. R.). IX-Liviana Press, Padova.

Boorstein, S. M. and Ewald, P. W. (1987) Costs and benefits of behavioral fever in *Melanoplus sanguinipes* infected by *Nosema acridophagus*. *Physiol. Zool*. **60**: 586–95.

Bradshaw, S. D. (1986) *Ecophysiology of Desert Reptiles*. Academic Press, Sydney.

Bradshaw, S. D., Cohen, D., Katsaros, A., Tom, J. and Owen, F. J. (1987) Determination of ^{18}O by prompt nuclear reaction analysis: application for measurement of microsamples. *J. Appl. Physiol*. **63**: 1296–302.

Brafield, A. E. and Llewellyn, M. J. (1982) *Animal Energetics*. Blackie, London.

Brain, C. (1991) Activity of water-deprived baboons (*Papio ursinus*) during inter-troop encounters. *S. Afr. J. Sci*. **84**: 590–1.

Braunitzer, G. and Hiebl, I. (1988) Molekulare aspekte der höhenatmung von vögeln. *Naturwissenschaften* **75**: 280–7.

Briegel, H. (1990) Fecundity, metabolism, and body size in *Anopheles* (Diptera: Culicidu), vectors of malaria. *J. Med. Entomol*. **27**: 839–50.

Broza, M. (1979) Dew, fog and hygroscopic food as a source of water for desert arthropods. *J. Arid Environ*. **2**: 43–9.

Buffenstein, R. and Louw, G. N. (1982) Temperature effects on bioenergetics of growth, assimilation efficiency and thyroid activity in juvenile varanid lizards. *J. Thermal Biol*. **7**: 197–200.

Burrage, B. R. (1973) Comparative ecology and behaviour of *Chamaeleo pumilus pumilus* (Gmelin) and *C. namaquensis* A. Smith (Sauria: chamaeleonidae). *Ann. S. Afr. Mus*. **61**: 1–158.

Cabanac, M. (1989) Fever in the leech, *Nephelopsis obscura* (Annelida). *J. Comp. Physiol. B* **159**: 281–5.

Cade, T. T. and MacLean, G. L. (1976) Transport of water by adult sandgrouse to their young. *Condor* **69**: 323–43.

Callow, P. (1978) *Life Cycles*. Chapman and Hall, London.

Carey, F. G. and Teal, J. M. (1966) Heat conservation in tuna fish muscle. *Proc. Natl. Acad. Sci*. **56**: 1464–9.

Chappell, M. A. and Bartholomew, G. A. (1981) Standard operative temperatures and thermal energetics of the antelope ground squirrel *Ammospermophilus Leucurus*. *Physiol. Zool*. **54**: 81–93.

Childress, J. J. (1987) Uptake and transport of sulfide in marine invertebrates. In: *Comparative Physiology: Life in Water and on Land* (eds Dejours, P., Bolis, L., Taylor, C. R. and Weibel, E. R.). IX-Liviana Press, Padova.

Choshniak, I. and Shkolnik, A. (1978) The rumen as a protective osmotic mechanism during rapid rehydration in the black Bedouin goat. In: *Osmotic and Volume Regulation*. Alfred Benzon Symposium XI, Munksgaard.

Clegg, J. S. (1964) The control of emergence and metabolism by external osmotic pressure and the role of free glycerol in developing cysts of *Artemia salina. J. Exp. Biol.* **41**: 879–92.

Clegg, J. S. (1987) Cellular and molecular adaptations to severe water loss. In: *Comparative Physiology: Life in Water and on Land* (eds Dejours, P., Bolis, L., Taylor, C. R. and Weibel, E. R.). IX-Liviana Press, Padova.

Clutton-Brock, T. H. (1988) *Reproductive Success.* University of Chicago Press, Chicago.

Colinvaux, P. (1978) *Why Big Fierce Animals Are Rare.* George Allen and Unwin, Boston.

Couthié, P. A. and Machin, J. (1984) Allometry of water vapor absorption in two species of tenebrionid beetle larvae. *Am. J. Physiol.* **247**: R230–R236.

Crew, F. A. E. (1922) A suggestion as to the cause of the aspermatic condition of the imperfectly descended testis. *J. Anat.* **56**: 98.

Crews, D. and Garstaka, W. R. (1982) The ecological physiology of a garter snake. *Sci. Amer.* **247**: 136–44.

Dawkins, M. J. R. and Hull, D. (1965) The production of heat by fat. *Sci. Amer.* **213**: 62–7.

Dawson, T. J., Robertshaw, D. and Taylor, C. R. (1974) Sweating in the kangaroos: a cooling mechanism during exercise, but not in the heat. *Am. J. Physiol.* **227**: 494–8.

Dawson, W. R., Shoemaker, V. H. and Licht, P. (1966) Evaporative water losses in some small Australian lizards. *Ecology* **47**: 589–94.

Dejours, P. (1989) From comparative physiology of respiration to several problems of environmental adaptations and to evolution. *J. Physiol.* **410**: 1–19.

Diamond, J. M. and Buddington, R. K. (1987) Intestinal nutrient absorption in herbivores and carnivores. In: *Comparative Physiology: Life in Water and on Land* (eds Dejours, P., Bolis, L., Taylor, C. R. and Weibel, E. R.). IX-Liviana Press, Padova.

Diamond, J. M., Karasov, W. H., Phan, D. and Carpenter, F. L. (1986) Digestive physiology is a determinant of foraging bout frequency in hummingbirds. *Nature* **320**: 62–3.

Dixon, J. E. W. and Louw, G. N. (1978) Seasonal effects on nutrition, reproduction and aspects of thermoregulation in the Namaqua sandgrouse (*Pterocles namaqua*). *Madoqua* **11**: 19–29.

Dobson, A. P. and Crawley, M. J. (1987) What's special about desert ecology? *Trends Ecol. Evol.* **2**: 145–6.

Durnin, J. V. G. A. and Passmore, R. (1967) *Energy, Work and Leisure.* Heinemann, London.

Dussourd, D. E. and Eisner, T. (1987) Vein-cutting behaviour: insect counterploy to the latex defense of plants. *Science* **237**: 898–901.

Eckert, R. and Randall, D. J. (1983) *Animal Physiology: Mechanisms and Adaptations.* Freeman, San Francisco.

Edgar, J. A. and Culvenor, C. C. J. (1974) Pyrrolizidine ester alkaloid in danaid butterflies. *Nature* **248**: 614–16.

Edney, E. B. (1977) *Water Balance in Land Arthropods.* Springer-Verlag, Berlin.

Epting, R. J. (1980) Functional dependence of the power for hovering on wing disc loading in hummingbirds. *Physiol. Zool.* **53**: 347–57.

Feder, M. E., Bennett, A. F., Burggren, W. W. and Huey, R. B. (eds) (1987) *New Directions in Ecological Physiology.* Cambridge University Press, Cambridge.

Feldman, H. A. and McMahon, T. A. (1983) The 3/4 mass exponent for energy metabolism is not a statistical artifact. *Resp. Physiol.* **52**: 149–63.

Fielden, L. J., Waggoner, J. P., Perrin, M. R. and Hickman, G. C. (1990) Thermoregulation in the Namib Desert golden mole, *Eremitalpa granti namibensis* (Chrysochloridae). *J. Arid Environ.* **18**: 221–37.

Folk, G. E. (1974) *Textbook of Environmental Physiology*, 2nd edn. Lea and Febiger, Philadelphia.

Fyhn, H. J. (1979) Rodents. In: *Comparative Physiology of Osmoregulation in Animals*, Vol. II (ed. Maloiy, G. M. O.). Academic Press, London.

Gagge, A. P. (1940) Standard operative temperature: a generalised temperature scale applicable to direct and partitional calorimetry. *Am. J. Physiol.* **131**: 93–103.

Goldsworthy, G. J., Robinson, J. and Mordue, W. (1981) *Endocrinology*. Blackie, London.

Gordon, M. S. (1982) Water and solute metabolism. In: *Animal Physiology. Principles and Adaptations* (ed. Gordon, M. S.). Macmillan, New York.

Gould, S. J. and Lewontin, R. C. (1979) The spandrels of San Marco and the Panglossian paradigm: a critique of the adaptionist programme. *Proc. R. Soc. Lond. B* **205**: 581–98.

Graham, J. B. (1990) Ecological, evolutionary and physical factors influencing aquatic animal respiration. *Am. Zool.* **30**: 137–46.

Greenwald, L. (1989) Urine concentrating ability, body size and metabolic rate. *Am. Zool.* **29**: 127a.

Grojean, R. E., Sousa, J. A. and Henry, M. C. (1980) Utilization of solar radiation by polar animals: an optical model for pelts. *Appl. Optics* **19**: 339–46.

Grubb, P. J. and Whittaker, J. B. (1988) *Towards a More Exact Ecology*. Blackwell, Oxford.

Hadley, N. F. (1972) Desert species and adaptation. *Am. Sci.* **60**: 338–47.

Hadley, N. F. (1977) Epicuticular lipids of the desert tenebrionid beetle, *Eleodes armata*: seasonal and acclimatory effects on composition. *Insect Biochem.* **7**: 277–83.

Hadley, N. F. (1979) Wax secretion and color phases of the desert tenebrionid beetle *Cryptoglossa verrucosa* (Le Conte). *Science* **203**: 367–9.

Hadley, N. F. (1985) *The Adaptive Role of Lipids in Biological Systems*. John Wiley, New York.

Hadley, N. F. (1988) Zoology Department, Arizona, State University, Tempe, Arizona. Personal communication.

Hadley, N. F. (1989) Lipid water barriers in biological systems. *Prog. Lipid Res.* **28**: 1–33.

Hadley, N. F. and Louw, G. N. (1980) Cuticular hydrocarbons and evaporative water loss in two tenebrionid beetles from the Namib Desert. *S. Afr. J. Sci.* **76**: 298–301.

Hadley, N. F., Toolson, E. C. and Quinlan, M. C. (1989) Regional differences in cuticular permeability in the desert cicada *Diceroprocta apache*: implications for evaporative cooling. *J. Exp. Biol.* **141**: 219–30.

Hagedorn, H. H. (1983) The role of ecdysteroids in the adult insect. In: *Endocrinology of Insects*, Vol. 1 (eds Downer, R. G. H. and Laufer, H.). Alan R. Liss Inc., New York.

Hales, J. R. S. (1966) The partition of respiratory ventilation of the panting ox. *J. Physiol.* **188**: 45–68.

Hamamura, Y., Hayashiya, K., Naito, K., Matasuura, K. and Nishida, J. (1962) Food selection by silkworm larvae. *Nature* **194**: 754–5.

Hamilton, W. J. III (1975) Coloration and its thermal consequences for diurnal desert insects. In: *Environmental Physiology of Desert Organisms* (ed. Hadley, N. F.). Dowden, Hutchinson and Ross, Stroudsburg, Pennsylvania.

Hamilton, W. J. III and Seely, M. K. (1976) Fog basking by the Namib Desert beetle, *Onymacris unguicularis*. *Nature* **262**: 284–5.

Harborne, J. B. (1982) *Introduction to Ecological Biochemistry*. Academic Press, London.

Heinrich, B. (1976) Heat exchange in relation to blood flow between thorax and abdomen in bumblebees. *J. Exp. Biol.* **64**: 561–85.

Heinrich, B. and Mommsen, T. B. (1985) Flight of winter moths near 0 °C. *Science* **228**: 177–9.

Hellon, R. F. (1967) Thermal stimulation of hypothalamic neurons in unanaesthetized rabbits. *J. Physiol.* **193**: 381–95.

Helversen, O. V. and Reyer, H.-U. (1984) Nectar intake and energy expenditure in a flower visiting bat. *Oecologia* **63**: 178–84.

Hemmingsen, A. M. (1960) Energy metabolism as related to body size and respiratory surfaces, and its evolution. *Rep. Steno Mem. Hosp. (Copenhagen)* **9**: 1–110.

Henschel, J. R. (1990) Spiders wheel to escape. *S. Afr. J. Sci.* **86**: 151–2.

Henschel, J. R. and Lubin, Y. D. (1990) Foraging at the thermal limit: burrowing spiders (Eresidae) in the Namib Desert dunes. *Oecologia* **84**: 461–7.

Heusner, A. A. (1982) Energy metabolism and body size. I. Is the 0.75 mass exponent of Kleiber's equation a statistical artifact? *Resp. Physiol.* **48**: 1–12.

Hochachka, P. W. (1980) *Living Without Oxygen: Closed and Open Systems in Hypoxia Tolerance*. Harvard University Press, Cambridge, Mass.

Hochachka, P. W. and Somero, G. N. (1973) *Strategies of Biochemical Adaptation*. W. B. Saunders, Philadelphia.

Hofmann, R. R. (1989) Evolutionary steps of ecophysiological adaption and diversification of ruminants: a comparative view of their digestive system. *Oecologia* **78**: 443–57.

Hofmeyr, H. S., Guidry, A. J. and Waltz, F. A. (1969) Effects of temperature and wool length on surface and respiratory evaporative losses of sheep. *J. Appl. Physiol.* **26**: 517–23.

Hofmeyr, M. D. and Louw, G. N. (1987) Thermoregulation, pelage conductance and renal function in the desert-adapted springbok, *Antidorcas marsupialis*. *J. Arid Environ.* **13**: 137–51.

Huey, R. B. (1982) Temperature, physiology and the ecology of reptiles. In: *Biology of the Reptilia*, Vol. 12 (eds Gans, G. and Pough, F. H.). Academic Press, New York.

Huey, R. B., Peterson, C. R., Arnold, S. J. and Porter, W. P. (1989) Hot rocks and not-so-hot rocks: retreat-site selection by garter snakes and its thermal consequences. *Ecology*, **70**: 931–44.

Isack, H. A. and Reyer, H.-U. (1989) Honeyguides and honeygatherers: Interspecific communication in a symbiotic relationship. *Science* **243**: 1343–6.

Janis, C. (1976) The evolutionary strategy of the Equidae and the origins of rumen and cecal digestion. *Evolution* **30**: 757–74.

Janzen, D. H. (1975) *Ecology of Plants in the Tropics* (Studies in Biology No. 58). Edward Arnold, London.

Jarvis, J. U. M. (1978) Energetics of survival in *Heterocephalus glaber* (Rüppell), the naked mole rat (Rodentia: Bathyergidae). *Bull. Carnegie Mus. Nat. Hist.* No. 6.

Johnsen, H. K., Blix, A. S., Jorgensen, L. and Mercer, J. B. (1985) Vascular basis for regulation of nasal heat exchange in reindeer. *Am. J. Physiol.* **249**: R617–R623.

Kanatani, H. (1969) Induction of spawning and oocyte maturation by 1-methyladenine in starfishes. *Exp. Cell Res.* **57**: 333–7.

Kanatani, H. and Nagahama, Y. (1983) Echinodermata. In: *Reproductive Biology of Invertebrates*, Vol. 1 (eds Adiyode, K. G. and Adiyodi, R. G.). John Wiley, New York.

Karasov, W. H. (1987) Nutrient requirements and the design and function of guts in fish, reptiles and mammals. In: *Comparative Physiology: Life in Water and on Land* (eds Dejours, P., Bolis, L., Taylor, C. R. and Weibel, E. R.). IX-Liviana Press, Padova.

Kasting, N. W., Adderley, S. A. L., Safford, T. and Hewlett, K. G. (1989) Thermoregulation in Beluga (*Delphinapterus leucas*) and killer (*Orcinus orca*) whales. *Physiol. Zool.* **62**: 687–701.

Katz, U. (1973) The effects of water deprivation and hypertonic salt injection on several rodent species compared with the albino rat. *Comp. Biochem. Physiol.* **44A**: 473–85.

Kay, D. G. and Brafield, A. E. (1973) The energy relations of polychaete the *Neanthes (=Nereis) virens* (Sars). *J. Anim. Ecol.* **42**: 673–92.

Keeler, R. F. (1975) Toxins and teratogens of higher plants. *Lloydia* **38**: 56–86.

Kilgore, D. L., Boggs, D. F. and Birchard, G. F. (1979) Role of the rete mirabile ophthalmicum in maintaining the body to brain temperature difference in pigeons. *J. Comp. Physiol.* **129**: 119–22.

Kleiber, M. (1932) Body size and metabolism. *Hilgardia* **6**: 315–53.

Klowden, M. C. and Blackmer, J. L. (1987) Humoral control of pre-oviposition behaviour in the mosquito, *Aedes aegypti*. *J. Insect Physiol.* **33**: 689–92.

Kluger, M. J. (1980) Fever. *Pediatrics* **66**: 720–4.

Kluger, M. J., Ringler, D. H. and Anver, M. R. (1975) Fever and survival. *Science* **188**: 166–8.

Kobelt, F. and Linsenmair, K. E. (1986) Adaptations of the reed frog *Hyperolius viridiflavus* (Amphibia, Anura, Hyperoliidae) to its arid environment. I. The skin of *Hyperolius viridiflavus nitidulus* in wet and dry season conditions. *Oecologia* **68**: 533–41.

Koehn, R. K. (1984) *The Application of Genetics to Problems of the Marine Environment: Future Areas of Research*. Natural Environment Research Council, Swindon, U.K.

Koehn, R. K. (1987) The importance of genetics to physiological ecology. In: *New Directions in Ecological Physiology* (eds Feder, M. E., Bennett, A. E., Burggren, W. W. and Huey, R. B.). Cambridge University Press, Cambridge.

Koehn, R. K. and Immerman, F. W. (1981) Biochemical studies of aminopeptidase polymorphism in *Mytilus edulis* I: Dependence of enzyme activity on season, tissue and genotype. *Biochem. Genet.* **19**: 1115–42.

Koehn, R. K. and Siebenaller, J. F. (1981) Biochemical studies of aminopeptidase polymorphism in *Mytilus edulis* II. Dependence of reaction rate on physical factors and enzyme concentration. *Biochem. Genet.* **19**: 1143–62.

Kooyman, G. L. (1981) *Weddell Seal: Consummate Diver*. Cambridge University Press, New York.

Krebs, J. R. and Davies, N. B. (1981) *An Introduction to Behavioural Ecology*. Blackwell, Oxford.

Laburn, H. P., Mitchell, D., Mitchell, G. and Saffy, K. (1988) Effects of tracheostomy breathing on brain and body temperatures in hyperthermic sheep. *J. Physiol.* **406**: 331–44.

Lack, D. (1948) The significance of litter size. *J. Anim. Ecol.* **17**: 45–50.

Licht, P. (1964) The temperature dependence of myosin-adenosinetriphosphatase and alkaline phosphatase in lizards. *Comp. Biochem. Physiol.* **12**: 331–40.

Linsenmair, K. E. (1987) Kin recognition in subsocial arthropods, in particular in the desert isopod *Hemilepistus reaumuri*. In: *Kin Recognition in Animals* (eds Fletcher, D. J. C. and Michener, C. D.). John Wiley, New York.

Linthicum, D. S. and Carey, F. G. (1972) Regulation of brain and eye temperatures by the bluefin tuna. *Comp. Biochem. Physiol.* **43**: 425–33.

Lombard, A. T. (1989) Unpublished Ph.D. Thesis, University of Cape Town.

Louw, G. N. (1984) Water deprivation in herbivores under arid conditions. In: *Herbivore Nutrition in the Tropics and Sub-Tropics* (ed. Gilchrist, F.). Science Press, Craighall, Johannesburg.

Louw, G. N. and Hadley, N. F. (1985) Water economy of the honeybee: a stoichiometric accounting. *J. Exp. Zool.* **235**: 147–50.

Louw, G. N. and Nicolson, S. W. (1983) Thermal, energetic and nutritional considerations in the foraging and reproduction of the carpenter bee *Xylocopa capitata*. *J. Ent. Soc. S. Afr.* **46**: 227–40.

Louw, G. N. and Seely, M. K. (1982) *Ecology of Desert Organisms*. Longman, London.

Louw, G. N., Belonje, P. C. and Coetzee, H. J. (1969) Renal function, respiration, heart rate and thermoregulation in the ostrich (*Struthio camelus*). *Scient. Pap. Namib Desert Res. Stat.* **42**: 43–54.

Louw, G. N., Young, B. A. and Bligh, J. (1976) Effect of thyroxine and noradrenaline on thermoregulation, cardiac rate and oxygen consumption in the monitor lizard *Varanus albigularis albigularis*. *J. Thermal Biol.* **1**: 189–93.

Louw, G. N., Nicolson, S. W. and Seely, M. K. (1986) Respiration beneath desert sand: carbon dioxide diffusion and respiratory patterns in a tenebrionid beetle. *J. Exp. Biol.* **120**: 443–7.

Lovegrove, B. G. (1986) The metabolism of social subterranean rodents: adaptation to aridity. *Oecologia* **69**: 551–5.

Lovegrove, B. G. and Painting, S. (1987) Variations in the foraging and burrow structures of the Damara molerat *Cryptomys damarensis* in the Kalahari Gemsbok National Park. *Koedoe* **30**: 149–63.

Lovegrove, B. G. and Wissel, C. (1988) Sociality in molerats, metabolic scaling and the role of risk sensitivity. *Oecologia* **74**: 600–6.

Loveridge, J. P. (1970) Observations on nitrogenous excretion and water relations of *Chiromantis xerampelina* (Amphibia, Anura). *Arnoldia* **5**: 1–6.

Loveridge, J. P. (1975) Studies on the water balance of adult locusts. III. The water balance of non-flying locusts. *Zoologica Africana* **10**: 1–28.

Lowry, W. P. (1969) *Weather and Life*. Academic Press, New York.

MacArthur, R. H. and Wilson, E. O. (1967) *The Theory of Island Biogeography*. Princeton University Press, Princeton, NJ.

MacFarlane, W. V., Morris, R. J. H., Howard, B. and Budtz-Olsen, O. E. (1959) Extracellular fluid distribution in tropical Merino sheep. *Aust. J. Agric. Res.* **10**: 269–86.

McArthur, A. J. and Clark, J. A. (1987) Body temperature and heat and water balance. *Nature* **326**: 647–8.

McClain, E., Seely, M. K., Hadley, N. F. and Gray, V. (1985) Wax blooms in tenebrionid beetles of the Namib Desert: correlations with environment. *Ecology* **66**: 112–18.

McClanahan, L. L., Stinner, J. N. and Shoemaker, V. H. (1978) Skin lipids, water loss and energy metabolism in a South American tree frog (*Phyllomedusa saugagei*). *Physiol. Zool.* **51**: 179–87.

McFadden, E. R. (1983) Respiratory heat and water exchange: physiological and clinical implications. *J. Appl. Physiol.* **54**: 331–6.

Machin, J. (1981) Water compartmentalisation in insects. *J. Exp. Zool.* **215**: 327–33.

Maddrell, S. (1987) Osmoregulation in terrestrial and aquatic insects. In: *Comparative Physiology: Life in Water and on Land* (eds Dejours, P., Bolis, L., Taylor, C. R. and Weibel, E. R.). IX-Liviana Press, Padova.

Maloiy, G. M. O., MacFarlane, W. V. and Shkolnik, A. (1979) Mammalian herbivores. In: *Comparative Physiology of Osmoregulation in Animals*, Vol. 2 (ed. Maloiy, G. M. O.). Academic Press, London.

Marsh, A. C. (1985) Thermal responses and temperature tolerance in a diurnal desert ant, *Ocymyrmex barbiger*. *Physiol. Zool.* **58**: 629–36.

Marshall, F. H. A. (1965) The breeding season. In: *Marshall's Physiology of Reproduction*, Vol. 1 (ed. Parkes, A. S.). Longman, London.

Martin, R. D. (1980) Body temperature, activity and energy costs. *Nature* **283**: 335–6.

Millar, R. P. (1972) Reproduction in the rock hyrax (*Procavia capensis*) with special references to seasonal sexual activity in the male. Ph.D. Thesis, University of Liverpool.

Millar, R. P. and King, J. A. (1988) Evolution of gonadotropin-releasing hormone: multiple usage of a peptide. *NIPS, Int. Union Physiol. Sci./Am. Physiol. Soc.* **3**: 49–53.

Mitchell, D. (1974) Convective heat loss from man and other animals. In: *Heat Loss from Animals and Man* (eds Monteith, J. L. and Mount, L. E.). Butterworths, London.

Mitchell, D. and Laburn, H. P. (1985) Pathophysiology of temperature regulation. *Physiologist* **28**: 507–17.

Mitchell, D., Laburn, H. P., Nijland, M. J. M., Zurovsky, Y. and Mitchell, G. (1987) Selective brain cooling and survival. *S. Afr. J. Sci.* **83**: 598–604.

Mitchell, D., Laburn, H. P., Matter, M. and McClain, E. (1990) Fever in Namib and other ectotherms. *Transvaal Museum Monograph* **7**: 179–92.

Moen, A. N. (1973) *Wildlife Ecology*. Freeman, San Francisco.

Mugaas, J. N. and King, J. R. (1981) Annual variation of daily energy expenditure by the black-billed magpie. *Studies in Avian Biology* No. 5. Cooper Ornithological Society.

Nagy, K. A. (1980) CO_2 production in animals: analysis of potential errors in the doubly labeled water method. *Am. J. Physiol.* **238**: R466–R473.

Nagy, K. A. (1983) *The Doubly Labeled Water ($^3HH^{18}O$) Method: A Guide to its Use.* University of California, Los Angeles, Publication No. 12–1417.

Nagy, K. A. (1987) Field metabolic rate and food requirement scaling in mammals and birds. *Ecol. Monogr.* **57**: 111–28.

Nagy, K. A. and Medica, P. A. (1986) Physiological ecology of desert tortoises in southern Nevada. *Herpetologica* **42**: 73–92.

Nagy, K. A. and Peterson, C. C. (1987) Water flux scaling. In: *Comparative Physiology: Life in Water and on Land* (eds Dejours, P., Bolis, L., Taylor, C. R. and Weiberl, E. R.). IX-Liviana Press, Padova.

Newell, R. C. (1979) *Biology of Intertidal Animals*, 3rd edn. Marine Ecology Surveys, Faversham, Kent.

Nicolson, S. W. (1980) Water balance and osmoregulation in *Onymacris plana*, a tenebrionid beetle from the Namib Desert. *J. Insect Physiol.* **26**: 315–20.

Nicolson, S. W. (1990) Water relations of the Namib tenebrionid beetles. *Transvaal Museum Monograph* No. 7: 173–8.

Nicolson, S. W. and Louw, G. N. (1982) Simultaneous measurement of evaporative water loss, oxygen consumption and thoracic temperature during flight in a carpenter bee. *J. Exp. Zool.* **222**: 287–96.

O'Connor, R. J. (1984) *The Growth and Development of Birds.* John Wiley and Sons, New York.

O'Donnel, M. J. (1978) The site of water vapour absorption in *Arenivaga investigata*. In: *Comparative Physiology: Water, Ions and Fluid Mechanics* (eds Schmidt-Nielsen, K., Bolis, L. and Maddrell, S. H. P.). Cambridge University Press, Cambridge.

O'Donnell, M. J. (1987) Water vapour absorption by arthropods: different sites, different mechanisms. In: *Comparative Physiology: Life in Water and on Land* (eds Dejours, P., Bolis, L., Taylor, C. R. and Weibel, E. R.). IX-Liviana Press, Padova.

Øritsland, N. A. and Ronald, K. (1978) Solar heating of mammals: observations of hair transmittance. *Int. J. Biometeor.* **22**: 197–201.

Paladino, F. V. and King, J. R. (1984) Thermoregulation and oxygen consumption during terrestrial locomotion by white-crowned sparrows *Zonotrichia leucophrys gambelii*. *Physiol. Zool.* **57**: 226–36.

Pearson, O. P. (1950) The metabolism of hummingbirds. *Condor* **52**: 145–52.

Pearson, O. P. (1954) Habits of the lizard *Liolaemus multiformis multiformis* at high altitudes in southern Peru. *Copeia* 1954: 111–16.

Peters, R. H. (1983) *The Ecological Implications of Body Size.* Cambridge University Press, Cambridge.

Pettenkofer, M. and Voit, C. (1866) Untersuchungen über den Stoffverbrauch des normalen Menschen. *Z. Biol.* **2**: 459–573.

Phillips, J. G., Butler, P. J. and Sharp, P. J. (1985) *Physiological Strategies in Avian Biology.* Blackie, London.

Pianka, E. R. (1970) On *r*- and *k*-selection. *Am. Nat.* **104**: 592–7.

Pianka, E. R. (1986) *Ecology and Natural History of Desert Lizards.* Princeton University Press, Princeton, NJ.

Pietruszka, R. D., Hanrahan, S. A., Mitchell, D. and Seely, M. K. (1986) Lizard herbivory in a sand dune environment: the diet of *Angolosaurus skoogi*. *Oecologia* **70**: 587–91.

Porter, W. P. and Gates, D. M. (1969) Thermodynamic equilibria of animals with environment. *Ecol. Monogr.* **39**: 227–45.

Pough, F. H. and Andrews, R. M. (1985) Use of anaerobic metabolism by free-ranging lizards. *Physiol. Zool.* **58**: 205–13.

Prosser, C. L. (ed.) (1973) *Comparative Animal Physiology.* W. B. Saunders, Philadelphia.

Rankin, J. C. and Davenport, J. (1981) *Animal Osmoregulation.* Blackie, Glasgow.

Rasa, O. A. E. (1990) Evidence for subsociality and division of labour in a desert tenebrionid beetle *Parastizopus armaticeps* Peringuey. *Die Naturwissenschaften* **76**: 273–4.

Raubenheimer, D. and Simpson, S. J. (1990) The effects of simultaneous variation in protein, digestible carbohydrate and tannic acid on the feeding behaviour of larval *Locusta migratoria* (L.) and *Schistocerca gregaria* (Forskal). I. Short-term studies. *Physiol. Entomol.* **15**: 219–33.

Rehr, S. S., Feeny, P. P. and Janzen, D. H. (1973) Chemical defense in Central American non-ant-acacias. *J. Anim. Ecol.* **42**: 405–16.

Reyer, H.-U. (1980) Flexible helper structure as an ecological adaptation in the pied kingfisher (*Ceryle rudis rudis.* L.). *Behav. Ecol. Sociobiol.* **6**: 219–27.

Reyer, H.-U. and Westerterp, K. (1985) Parental energy expenditure: a proximate cause of helper recruitment in the pied kingfisher (*Ceryle rudis*). *Behav. Ecol. Sociobiol.* **17**: 363–9.

Reyer, H.-U., Dittami, J. P. and Hall, M. R. (1986) Avian helpers at the nest: are they psychologically castrated? *Ethology* **71**: 216–28.

Richards, S. A. (1973) *Temperature Regulation.* Wykeham Publications, London.

Ricklefs, R. E. (1974) Energetics of reproduction in birds. In: *Avian Energetics* (ed. Paynter, R. A.). Nuttal Ornithological Club, Cambridge, Mass.

Schartz, R. L. and Zimmerman, J. L. (1971) The time and energy budget of the male dickcissel (*Spiza americana*). *Condor* **73**: 65–76.

Schmidt-Nielsen, K. (1964) *Desert Animals: Physiological Problems of Heat and Water.* Oxford University Press, New York.

Schmidt-Nielsen, K. (1972) *How Animals Work.* Cambridge University Press, London.

Schmidt-Nielsen, K. (1983) *Animal Physiology: Adaptation and Environment,* 3rd edn. Cambridge University Press, Cambridge.

Schmidt-Nielsen, K. (1984) *Scaling, Why is Animal Size So Important?* Cambridge University Press, Cambridge.

Schmidt-Nielsen, B. and Schmidt-Nielsen, K. (1951) A complete account of the water metabolism in kangaroo rats and an experimental verification. *J. Cell. Comp. Physiol.* **38**: 165–81.

Schmidt-Nielsen, K., Schmidt-Nielsen, B., Jarnum, S. A. and Houpt, J. R. (1957) Body temperature of the camel and its relation to water economy. *Am. J. Physiol.* **188**: 103–12.

Schmidt-Nielsen, K., Schroter, R. C. and Shkolnik, A. (1981) Desaturation of the exhaled air in the camel. *Proc. Roy. Soc. B* **211**: 305–19.

Schmuck, R., Kobelt, F. and Linsenmair, K. E. (1988) Adaptations of the reed frog *Hyperolius viridiflavus* (Amphibia, Anura, Hyperoliidae) to its arid environment. *J. Comp. Physiol. B* **158**: 537–46.

Schoenmakers, H. J. N., Dieleman, S. J., Van Bohemen, Ch. G. and Voogt, P. A. (1978) Biosynthesis of steroids and their possible functions for reproduction in *Asterias rubens* (Echinodermata). In: *Comparative Endocrinology* (eds Gaillard, P. J. and Boer, H. H.). Elsevier, Amsterdam.

Scholander, P. F. (1955) Evolution of climatic adaptation in homeotherms. *Evolution* **9**: 15–26.

Scholander, P. F., Hock, R., Walters, V., Johnson, F. and Irving, L. (1950a) Heat regulation in some arctic and tropical mammals and birds. *Biol. Bull.* **99**: 237–58.

Scholander, P. F., Walters, V., Hock, R. and Irving, L. (1950b) Body insulation of some arctic and tropical mammals and birds. *Biol. Bull.* **99**: 225–36.

Seely, M. K. (1989) Desert invertebrate physiological ecology: what is special? *S. Afr. J. Sci.* **85**: 266–70.

Seely, M. K. and Hamilton, W. J. III (1976) Fog catchment sand trenches constructed by tenebrionid beetles, *Lepidochora*, from the Namib Desert. *Science* **193**: 484–6.

Seely, M. K., Roberts, C. S. and Mitchell, D. (1988) High body temperatures of Namib dune tenebrionids—why? *J. Arid. Environ.* **14**: 135–43.

Seely, M. K., Mitchell, D. and Goelst, K. (1990) Boundary layer microclimate and *Angolosaurus skoogi* (Sauria: Cordylidae) activity on a northern Namib dune. *Transvaal Museum Monograph* No. 7: 155–62.

Seibt, U. and Wickler, W. (1988) Bionomics and social structure of 'family spiders' of the genus *Stegodyphus*. *Verh. naturwiss. Ver. Hamburg.* **30**: 255–303.

Setchell, B. P. (1982) Spermatogenesis and spermatozoa. In: *Reproduction in Mammals*, 2nd edn. I. *Germ Cells and Fertilization* (eds Austin, C. R. and Short, R. J.). Cambridge University Press, Cambridge.

Sherry, D. F., Mrosovsky, N. and Hogan, J. A. (1980) Weight loss and anorexia during incubation in birds. *J. Comp. Physiol. Psychol.* **94**: 89–98.

Shkolnik, A. (1977) Physiological adaptations of mammals to life in the desert. In: *The Desert, Past, Present, Future* (ed. Ezra Sohar). University of Tel Aviv, Tel Aviv. (In Hebrew.)

Shkolnik, A., Borut, A., Choshniak, I. and Maltz, E. (1975) Water economy and drinking regime of the Bedouin goat. *Symposium Israel-France*, Volcani Center, Bet-Dagan, Israel, pp. 79–90.

Shoemaker, V. H. (1987) Osmoregulation in amphibians. In: *Comparative Physiology: Life in Water and on Land* (eds Dejours, P., Bolis, L., Taylor, C. R. and Weibel, E. R.). IX-Liviana Press, Padova.

Shoemaker, V. H., McClanahan, L. L., Withers, P. C., Hillman, S. S., Drewes, R. C. (1987) Thermoregulatory response to heat in the waterproof frogs *Phyllomedusa* and *Chiromantis*. *Physiol. Zool.* **60**: 365–72.

Short, R. V. (1984) Species differences in reproductive mechanisms. In: *Reproduction in Mammals*, Vol. 4, *Reproductive Fitness* (eds Austin, C. R. and Short, R. V.). Cambridge University Press, Cambridge.

Siegler, D. S. (1975) Isolation and characterization of naturally occurring cyanogenic compounds. *Phytochemistry* **14**: 9–30.

Silanikove, N., Tagari, H. and Shkolnik, A. (1980) Gross energy digestion and urea recycling in the desert black Bedouin goat. *Comp. Biochem. Physiol.* **67A**: 215–18.

Singh, P. and Moore, R. F. (eds) (1985) *Handbook of Insect Rearing*, Vols 1 and 2. Elsevier, Amsterdam.

Skadhauge, E. (1981) *Osmoregulation in Birds*. Springer-Verlag, Berlin.

Skinner, J. D. and Louw, G. N. (1966) Heat stress and spermatogenesis in *Bos indicus* and *Bos taurus* cattle. *J. Appl. Physiol.* **21**: 1784–90.

Skinner, J. D. and Van Jaarsveld, A. S. (1987) Adaptive significance of restricted breeding in southern African ruminants. *S. Afr. J. Sci.* **83**: 657–63.

Skinner, J. D., Van Aarde, R. J. and Van Jaarsveld, A. S. (1984) Adaptations in three species of large mammals. (*Antidorcas marsupialis, Hystrix africaeaustralis, Hyaena brunnea*), to arid environments. *S. Afr. J. Zool.* **19**: 82–6.

Somero, G. N. (1987) Organic osmolyte systems: convergent evolution in the design of the intracellular milieu. In: *Comparative Physiology: Life in Water and on Land* (eds Dejours, P., Bolis, L., Taylor, C. R. and Weibel, E. R.). IX-Liviana Press, Padova.

Southwick, E. E. (1985) Allometric relations, metabolism and heat conductance in clusters of honey bees at cool temperatures. *J. Comp. Physiol.* **156**: 143–9.

Stearns, S. C. (1987) *The Evolution of Sex and its Consequences*. Birkhäuser, Basel.

Steele, W. K. and Louw, G. N. (1988) Caecilians exhibit cutaneous respiration and high evaporative water loss. *S. Afr. J. Zool.* **23**: 134–5.

Taigen, T. L. and Pough, F. H. (1985) Metabolic correlates of anuran behaviour. *Am. Zool.* **25**: 987–97.

Taigen, T. L., Emerson, S. B. and Pough, F. H. (1982) Ecological correlates of anuran exercise physiology. *Oecologia* **52**: 49–56.

Taigen, T. L., Wells, K. D. and Marsh, R. L. (1985) The enzymatic basis of high metabolic rates in calling frogs. *Physiol. Zool.* **58**: 719–26.

Tallamy, D. W. (1985) Squash beetle feeding behaviour: an adaptation against induced cucurbit defenses. *Ecology* **66**: 1574–9.

Taylor, C. R. (1966) The vascularity and possible thermoregulatory function of the horns in goats. *Physiol. Zool.* **39**: 127–39.

Taylor, C. R. (1968) Hygroscopic food: a source of water for desert antelopes? *Nature* **219**: 181–2.

Taylor, C. R. (1969) The eland and the oryx. *Sci. Amer.* **220**: 88–95.

Taylor, C. R. (1970) Strategies of temperature regulation: effect on evaporation in East African ungulates. *Am. J. Physiol.* **219**: 1131–5.

Taylor, C. R. (1972) The desert gazelle: a paradox resolved. *Symp. Zool. Soc. Lond.* **31**: 215–27.

Taylor, C. R. (1987) Energetics of locomotion in water, on land and in air: what sets the cost? In: *Comparative Physiology: Life in Water and on Land* (eds Dejours, P., Bolis, L., Taylor, C. R. and Weibel, E. R.). IX-Liviana Press, Padova.

Taylor, C. R., Schmidt-Nielsen, K. and Raab, J. L. (1970) Scaling of energetic costs of running to body size in mammals. *Am. J. Physiol.* **219**: 1104–7.

Thomas, D. H. (1982) Salt and water excretion by birds: the lower intestine as an integrator of renal and intestinal excretion. *Comp. Biochem. Physiol.* **71A**: 527–36.

Thomas, L. (1985) Seven wonders. In: *The Sacred Beetle* (ed. Gardner, M.). Oxford University Press, Oxford.

Tracy, C. R. (1972) Newton's law: Its applicability for expressing heat losses from homeotherms. *BioScience* **22**: 656–9.

Tracy, C. R. (1976) A model of the dynamic exchanges of water and energy between a terrestrial amphibian and its environment. *Ecol. Monogr.* **46**: 293–326.

Treherne, J. (1987) Neural adaptations to osmotic and ionic stress in aquatic and terrestrial invertebrates. In: *Comparative Physiology: Life in Water and on Land* (eds Dejours, P., Bolis, L., Taylor, C. R. and Weibel, E. R.). IX-Liviana Press, Padova.

Tucker, V. A. (1975) The energetic cost of moving about. *Am. Sci.* **63**: 413–19.

Turner, J. S. (1988) Body size and thermal energetics: how should thermal conductance scale? *J. Thermal Biol.* **13**: 103–17.

Turner, J. S. and Tracy, C. R. (1985) Body size and the control of heat exchange in alligators. *J. Thermal Biol.* **10**: 9–11.

Tyler, M. J. (1985) Gastric brooding: a phenomenon unique to Australian frogs. *Search* **16**: 157–9.

Tyndale-Biscoe, C. H. (1982) What makes the kangaroo tick? CSIRO Information Sheet No. 1–38.

Tyndale-Biscoe, C. H. (1984) Mammalia–Marsupiala. In: *Marshall's Physiology of Reproduction*, 4th edn (ed. Lamming, G. E., Vol. I). Churchill Livingstone, London.

Unwin, D. M. (1980) *Microclimate Measurement for Ecologists*. Academic Press, London.

Vander Wall, S. B. and Balda, R. P. (1977) Coadaptations of the Clark's Nutcracker and the pinyon pine for efficient seed harvest and dispersal. *Ecol. Monogr.* **47**: 89–111.

Vaughn, L. K., Bernheim, H. A. and Kluger, M. J. (1974) Fever in the lizard *Dipsosaurus dorsalis*. *Nature* **252**: 473–4.

Veith, W. J. (1980) Viviparity and embryonic adaptations in the teleost *Clinus superciliosus*. *Can. J. Zool.* **58**: 1–12.

Vleck, C. M. and Vleck, D. (1987) Metabolism and energetics of avian embryos. *J. Exp. Zool.* (Suppl. 1): 111–25.

Vleck, D. (1979) The energy cost of burrowing by the pocket gopher *Thomomys bottae*. *Physiol. Zool.* **52**: 122–5.

Walsberg, G. E. (1983a) Coat color and solar heat gain in animals. *BioScience* **33**: 88–91.

Walsberg, G. E. (1983b) Avian ecological energetics. In: *Avian Biology*, Vol. 7 (eds Farner, D. S., King, J. R. and Parkes, K. C.). Academic Press, New York.

Walsberg, G. E., Campbell, G. S. and King, J. R. (1978) Animal coat colour and radiative heat gain: a re-evaluation. *J. Comp. Physiol.* **126**: 211–22.

Walton, M., Jayne, B. C. and Bennett, A. F. (1990) The energetic cost of limbless locomotion. *Science* **249**: 524–7.

Weast, R. C. (ed.) (1980) *CRC Handbook of Chemistry and Physics*, 60th edn. CRC Press Inc., Boca Raton, Florida.

Weathers, W. W. and Nagy, K. A. (1980) Simultaneous doubly labeled water ($^3HH^{18}O$) and time-budget estimates of daily energy expenditure in *Phainopepla nitens. Auk* **97**: 861–7.

Wehner, R. (1989) Strategien gegen den Hitzetod: Thermophilie und Thermoregulation bei Wüstenameisen (*Cataglyphis bombycina*). *Jubiläumsbd. Akad. Wiss. Lit. Mainz* 1989: 101–112.

Weibel, E. R., Taylor, C. R., Hoppeler, H. and Karas, R. H. (1987) Adaptive variation in the mammalian respiratory system in relation to energetic demand: I. Introduction to problem and strategy. *Resp. Physiol.* **69**: 1–6 and following papers.

Weis-Fogh, T. (1967) Respiration and tracheal ventilation in locusts and other flying insects. *J. Exp. Biol.* **47**: 561–87.

Whitford, W. G. (1973) The effects of temperature on respiration in the amphibia. *Am. Zool.* **13**: 505–12.

Willmer, P. G. (1982) Microclimate and the environmental physiology of insects. *Adv. Insect Physiol.* **16**: 1–57.

Willmer, P. G. and Unwin, D. (1981) Field analyses of insect heat budgets: reflectance, size and heating rates. *Oecologia* **50**: 250–5.

Wingerson, L. (1983) The lion, the spring and the pendulum. *New Scientist* 1983, **97**: 236–9.

Withers, D. C., Louw, G. N. and Henschel, J. (1980) Energetics and water relations of Namib Desert rodents. *S. Afr. J. Zool.* **15**: 131–7.

Withers, P. C. (1983) Energy, water, and solute balance of the ostrich *Struthio camelus*. *Physiol. Zool.* **56**: 568–79.

Withers, P. C., Siegfried, W. R. and Louw, G. N. (1981) Desert ostrich exhales unsaturated air. *S. Afr. J. Sci.* **77**: 569–70.

Withers, P., Louw, G. N. and Nicolson, S. (1982) Water loss, oxygen consumption and colour change in 'waterproof' reed frogs (*Hyperolius*). *S. Afr. J. Sci.* **78**: 30–2.

Wright, S. H. (1987) Nutrient transport across the integument of invertebrates. In: *Comparative Physiology: Life in Water and on Land* (eds Dejours, P., Bolis, L., Taylor, C. R. and Weibel, E. R.). IX-Liviana Press, Padova.

W-Worswick, P. V. (1987) Comparative study of colony thermoregulation in the African honeybee, *Apis mellifera adansonii* Latreille and the Cape honeybee, *Apis mellifera capensis* Escholtz. *Comp. Biochem. Physiol.* **86**: 95–102.

Index